Menachem Fisch

Library of Contemporary Jewish Philosophers

VOLUME 18

The titles published in this series are listed at brill.com/lcjp

Menachem Fisch

The Rationality of Religious Dispute

Edited by

Hava Tirosh-Samuelson and Aaron W. Hughes

BRILL

LEIDEN • BOSTON
2016

Cover Illustration: Courtesy of Uwe Dettmar.

The series *Library of Contemporary Jewish Philosophers* was generously supported by the Baron Foundation.

Library of Congress Cataloging-in-Publication Data

Names: Tirosh-Samuelson, Hava, 1950- editor. | Hughes, Aaron W., 1968- editor.
Title: Menachem Fisch : the rationality of religious dispute / edited by Hava Tirosh-Samuelson and Aaron W. Hughes.
Description: Leiden : Boston : Brill, 2016. | Series: Library of contemporary Jewish philosophers, ISSN 2213-6010 ; volume 18 | Includes bibliographical references.
Identifiers: LCCN 2016018305 (print) | LCCN 2016019799 (ebook) | ISBN 9789004323568 (hardback : alk. paper) | ISBN 9789004323582 (pbk. : alk. paper) | ISBN 9789004323575 (E-book)
Subjects: LCSH: Fisch, Menachem. | Science—Philosophy. | Jewish philosophy.
Classification: LCC B5800 .M46 2016 (print) | LCC B5800 (ebook) | DDC 181/.06—dc23
LC record available at https://lccn.loc.gov/2016018305

Typeface for the Latin, Greek, and Cyrillic scripts: "Brill". See and download: brill.com/brill-typeface.

ISSN 2213-6010
ISBN 978-90-04-32356-8 (hardback)
ISBN 978-90-04-32357-5 (e-book)

This hardback is also published in paperback under ISBN 978-90-04-32358-2.

This book is printed on acid-free paper and produced in a sustainable manner.

Printed by Printforce, the Netherlands

CONTENTS

THE CONTRIBUTORS

NOAH J. EFRON (Ph.D., Tel Aviv University, 1996) teaches at Bar-Ilan University, in the Graduate Program in Science, Technology, and Society, which he founded and chairs. He has been President of the Israeli Society for History and Philosophy of Science and is a member of the Executive Committee of the International Society for Science and Religion. Efron has been a member of the Institute for Advanced Study in Princeton, the Dibner Institute for History of Science and Technology at MIT, a Fellow at Harvard University, and a visiting professor at the Center for Bioethics of the University of Pennsylvania. He is the author, most recently, of *A Chosen Calling: Jews in Science in the Twentieth Century* (Johns Hopkins and HUC Presses, 2014) as well as other books and essays on the complicated intertwine of knowledge, religion, and politics. He played bass guitar for the band Liquid Plumr.

HAVA TIROSH-SAMUELSON (Ph.D., Hebrew University of Jerusalem, 1978) is Irving and Miriam Lowe Professor of Modern Judaism, the Director of Jewish Studies, and Professor of History at Arizona State University in Tempe, Arizona. Her research focuses on Jewish intellectual history, Judaism and ecology, science and religion, and feminist theory. In addition to numerous articles and book chapters in academic journals and edited volumes, she is the author of the award-winning *Between Worlds: The Life and Work of Rabbi David ben Judah Messer Leon* (SUNY Press, 1991) and the author of *Happiness in Premodern Judaism: Virtue, Knowledge, and Well-Being in Premodern Judaism* (Hebrew Union College Press, 2003). She is also the editor of *Judaism and Ecology: Created World and Revealed Word* (Harvard University Press, 2002); *Women and Gender in Jewish Philosophy* (Indiana University Press, 2004); *Judaism and the Phenomenon of Life: The Legacy of Hans Jonas* (Brill, 2008); *Building Better Humans? Refocusing the Debate on Transhumanism* (Peter Lang, 2011); *Hollywood's Chosen People: The Jewish Experience in American Cinema* (Wayne State University Press, 2012); and *Jewish Philosophy for the Twenty-First Century: Personal Reflections* (Brill, 2014). Professor Tirosh-Samuelson is the recipient of several large grants that have funded interdisciplinary research on religion, science, and technology.

Aaron W. Hughes (Ph.D., Indiana University Bloomington, 2000) holds the Philip S. Bernstein Chair in Jewish Studies at the University of Rochester. Hughes was educated at the University of Alberta, the Hebrew University of Jerusalem, and Oxford University. He has taught at Miami University of Ohio, McMaster University, the Hebrew University of Jerusalem, the University of Calgary, and the University at Buffalo. He is the author of over fifty articles and ten books, and the editor of seven books. His book titles include *Abrahamic Religions: On the Uses and Abuses of History* (Oxford University Press, 2012); *Muslim Identities* (Columbia University Press, 2013); *The Study of Judaism: Identity, Authenticity, Scholarship* (SUNY, 2013); and *Rethinking Jewish Philosophy: Beyond Particularism and Universalism* (Oxford University Press, 2014). He is also the Editor-in-Chief of *Method and Theory in the Study of Religion*.

EDITORS' INTRODUCTION TO THE SERIES

It is customary to begin studies devoted to the topic of Jewish philosophy by defining what exactly this term, concept, or even discipline is. We tend not to speak of Jewish mathematics, Jewish physics, or Jewish sociology, so why refer to something as "Jewish philosophy"? Indeed, this is the great paradox of Jewish philosophy. On the one hand it presumably names something that has to do with thinking, on the other it implies some sort of national, ethnic, or religious identity of those who engage in such activity. Is not philosophy just philosophy, regardless of who philosophizes? Why the need to append various racial, national, or religious adjectives to it?[1]

Jewish philosophy is indeed rooted in a paradox since it refers to philosophical activity carried out by those who call themselves Jews. As philosophy, this activity makes claims of universal validity, but as an activity by a well-defined group of people it is inherently particularistic. The question "What is Jewish philosophy?" therefore is inescapable, although over the centuries Jewish philosophers have given very different answers to it. For some, Jewish philosophy represents the relentless quest for truth. Although this truth itself may not be particularized, for such individuals, the use of the adjective "Jewish"—as a way to get at this truth—most decidedly is.[2] The Bible, the Mishnah, the Talmud, and related Jewish texts and genres are seen to provide particular insights into the more universal claims provided by the universal and totalizing gaze of philosophy. The problem is that these texts are not philosophical *on the surface*; they must, on the contrary, be interpreted to bring their philosophical insights to light. Within this context exegesis risks becoming eisegesis. Yet others eschew the term "philosophy" and instead envisage themselves as working

---

1 Alexander Altmann once remarked:

> It would be futile to attempt a presentation of Judaism as a philosophical system, or to speak of Jewish philosophy in the same sense as one speaks of American, English, French, or German philosophy. Judaism is a religion, and the truths it teaches are religious truths. They spring from the source of religious experience, not from pure reason.

See Alexander Altmann, "Judaism and World Philosophy," in *The Jews: Their History, Culture, and Religion*, ed. Louis Finkelstein (Philadelphia: Jewish Publication Society of America, 1949), vol. 2, 954.

2 In this regard, see Norbert M. Samuelson, *Jewish Faith and Modern Science: On the Death and Rebirth of Jewish Philosophy* (New York: Rowman and Littlefield, 2008), e.g., 10–12.

in a decidedly Jewish key in order to articulate or clarify particular issues that have direct bearing on Jewish life and existence.[3] Between these two perspectives or orientations, there exist several other related approaches to the topic of Jewish philosophy, which can and have included ethics,[4] gender studies,[5] multiculturalism,[6] and postmodernism.[7]

Despite their differences in theory and method, what these approaches have in common is that they all represent the complex intersection of Judaism, variously defined, and a set of non-Jewish grids or lenses used to interpret this rich tradition. Framed somewhat differently, Jewish philosophy—whatever it is, however it is defined, or whether definition is even possible—represents the collision of particularistic demands and universal concerns. The *universal*, or that which is, in theory, open and accessible to all regardless of race, color, creed, or gender confronts the *particular*, or that which represents the sole concern of a specific group that, by nature or definition, is insular and specific-minded.

Because it is concerned with a particular people, the Jews, and how to frame their traditions in a universal and universalizing light that is believed to conform to the dictates of reason, Jewish philosophy can never be about pure thinking, if indeed there ever can be such a phenomenon. Rather Jewish philosophy—from antiquity to the present—always seems to have had and, for the most part continues to have, rather specific and

---

[3] See, e.g., Strauss's claim about Maimonides' *Guide of the Perplexed*, perhaps one of the most important and successful works of something called Jewish philosophy ever written. He claims that one "begins to understand the *Guide* once one sees that it is not a philosophic book—a book written by a philosopher for philosophers—but a Jewish book: a book written by a Jew for Jews." See Leo Strauss, "How to Begin to Study *The Guide of the Perplexed*," in *The Guide of the Perplexed*, trans. Shlomo Pines, 2 vols. (Chicago: University of Chicago Press, 1963), vol. 1, xiv.

Modern iterations of this may be found, for example, in J. David Bleich, *Bioethical Dilemmas: A Jewish Perspective*, 2 vols. (vol. 1, New York: Ktav, 1998; vol. 2, New York: Targum Press, 2006).

[4] See, e.g., David Novak, *Natural Law in Judaism* (Cambridge: Cambridge University Press, 1998); Elliot Dorff, *Love Your Neighbor and Yourself: A Jewish Approach to Modern Personal Ethics* (New York: Jewish Publication Society of America, 2006).

[5] E.g., the collection of essays in *Women and Gender in Jewish Philosophy*, ed. Hava Tirosh-Samuelson (Bloomington, IN: Indiana University Press, 2004).

[6] E.g., Jonathan Sacks, *The Dignity of Difference: How to Avoid a Clash of Civilizations* (London: Continuum, 2003); Jonathan Sacks, *To Heal a Fractured World: The Ethics of Responsibility* (New York: Schocken, 2007).

[7] E.g., Elliot R. Wolfson, *Language, Eros, Being: Kabbalistic Hermeneutics and Poetic Imagination* (New York: Fordham University Press, 2004); Elliot R. Wolfson, *Open Secret: Postmessianic Messianism and the Mystical Revision of Menahem Mendel Schneerson* (New York: Columbia University Press, 2009); Elliot R. Wolfson, *A Dream Interpreted within a Dream: Oneiropoiesis and the Prism of Imagination* (New York: Zone Books, 2011).

perhaps even practical concerns in mind. This *usually* translates into the notion that Judaism—at least the Judaism that Jewish philosophy seeks to articulate—is comprehensible to non-Jews and, framed in our contemporary context, that Judaism has a seat at the table, as it were, when it comes to pressing concerns in the realms of ethics and bioethics.

Jewish philosophy, as should already be apparent, is not a disinterested subject matter. It is, on the contrary, heavily invested in matters of Jewish peoplehood and in articulating its aims and objectives. Because of this interest in concrete issues (e.g., ethics, bioethics, medical ethics, feminism) Jewish philosophy—especially contemporary Jewish philosophy—is often constructive as opposed to being simply reflective. Because of this, it would seem to resemble what is customarily called "theology" more than it does philosophy. If philosophy represents the critical and systematic approach to ascertain the truth of a proposition based on rational argumentation, theology is the systematic and rational study of religion and the articulation of the nature of religious truths. The difference between theology and philosophy resides in their object of study. If the latter has "truth," however we may define this term, as its primary object of focus, the former is concerned with ascertaining religious dogma and belief. They would seem to be, in other words, mutually exclusive endeavors.

What we are accustomed to call "Jewish philosophy," then, is a paradox since it does not—indeed, cannot—engage in truth independent of religious claims. As such, it is unwilling to undo the major claims of Judaism (e.g., covenant, chosenness, revelation), even if it may occasionally redefine such claims.[8] So although medieval Jewish thinkers may well gravitate toward the systematic thought of Aristotle and his Arab interpreters and although modern Jewish thinkers may be attracted to the thought of Kant and Heidegger, the ideas of such non-Jewish thinkers are always applied to Jewish ideas and values. Indeed, if they were not, those who engaged in such activities would largely cease to be Jewish philosophers and would instead become just philosophers who just happened to be Jewish (e.g., Henri Bergson, Edmund Husserl, and Karl Popper).

Whether in its medieval or modern guise, Jewish philosophy has a tendency to be less philosophical simply for the sake of rational analysis and more constructive. Many of the volumes that appear in the Library of

[8] A good example of what we have in mind here is the thought of Maimonides. Although he might well redefine the notion of prophecy, he never rejects the concept. On Maimonides on prophecy, see Howard Kreisel, *Prophecy: The History of an Idea in Medieval Jewish Philosophy* (Dordrecht: Kluwer, 2001), 148–56.

Contemporary Jewish Philosophers will bear this out. The truths of Judaism are upheld, albeit in often new and original ways. Although Jewish philosophy may well use non-Jewish ideas to articulate its claims, it never produces a vision that ends in the wholesale abandonment of Judaism.[9] Even though critics of Jewish philosophy may well argue that philosophy introduces "foreign" wisdom into the heart of Judaism, those we call Jewish philosophers do not perceive themselves to be tainting Judaism, but perfecting it or teasing out its originary meaning.[10]

The result is that Jewish philosophy is an attempt to produce a particular type of Judaism—one that is in tune with certain principles of rationalism. This rationalism, from the vantage point of the nineteenth century and up to the present, is believed to show Judaism in its best light, as the synthesis or nexus between a Greek-inflected universalism and the particularism of the Jewish tradition.

What is the status of philosophy among Jews in the modern period? Since their emancipation in the nineteenth century, Jews have gradually integrated into Western society and culture, including the academy. Ever since the academic study of Judaism began in the 1820s in Germany, Jewish philosophy has grown to become a distinctive academic discourse practiced by philosophers who now often hold positions in non-Jewish institutions of higher learning. The professionalization of Jewish philosophy has not been unproblematic, and Jewish philosophy has had to (and still has to) justify its legitimacy and validity. And even when Jewish philosophy is taught in Jewish institutions (for example, rabbinic seminaries or universities in Israel), it has to defend itself against those Jews who regard philosophy as alien to Judaism, or minimally, as secondary in importance to the inherently Jewish disciplines such as jurisprudence or exegesis. Jewish philosophy, in other words, must still confront the charge that it is not authentically Jewish.

The institutional setting for the practice of Jewish philosophy has shaped Jewish philosophy as an academic discourse. But regardless of the

---

9 This despite the claims of Yitzhak Baer who believed that philosophy had a negative influence on medieval Spanish Jews that made them more likely to convert to Christianity. See Israel Jacob Yuval, "Yitzhak Baer and the Search for Authentic Judaism," in *The Jewish Past Revisited: Reflections on Modern Jewish Historians*, ed. David N. Myers and David B. Ruderman (New Haven, CT: Yale University Press, 1998), 77–87.

10 Indeed, Jewish philosophers in the medieval period did not even see themselves as introducing foreign ideas into Judaism. Instead they saw philosophical activity as a reclamation of their birthright since the Jews originally developed philosophy before the Greeks and others stole it from them.

setting, Jewish philosophy as an academic discourse is quite distinct from Jewish philosophy as constructive theology, even though the two may often be produced by the same person.

Despite the lack of unanimity about the scope and methodology of Jewish philosophy, the Library of Contemporary Jewish Philosophers insists that Jewish philosophy has thrived in the past half century in ways that will probably seem surprising to most readers. When asked who are the Jewish philosophers of the twentieth century, most would certainly mention the obvious: Franz Rosenzweig (d. 1929), Martin Buber (d. 1965), and Emmanuel Levinas (d. 1995). Some would also be able to name Abraham Joshua Heschel (d. 1972), Mordecai Kaplan (d. 1983), Joseph Soloveitchik (d. 1993), and Hans Jonas (d. 1993). There is no doubt that these thinkers have either reshaped the discourse of Western thought for Jews and non-Jews or have inspired profound rethinking of modern Judaism. However, it is misleading to identify contemporary Jewish philosophy solely with these names, all of whom are now deceased.

In recent years it has been customary for Jews to think that Jewish philosophy has lost its creative edge or that Jewish philosophy is somehow profoundly irrelevant to Jewish life. Several reasons have given rise to this perception, not the least of which is, ironically enough, the very success of Jewish Studies as an academic discipline. Especially after 1967, Jewish Studies has blossomed in secular universities especially in the North American Diaspora, and Jewish philosophers have expressed their ideas in academic venues that have remained largely inaccessible to the public at large. Moreover, the fact that Jewish philosophers have used technical language and a certain way of argumentation has made their thought increasingly incomprehensible and therefore irrelevant to the public at large. At the same time that the Jewish public has had little interest in professional philosophy, the practitioners of philosophy (especially in the Anglo American departments of philosophy) have denied the philosophical merits of Jewish philosophy as too religious or too particularistic and excluded it entirely. The result is that Jewish philosophy is now largely generated by scholars who teach in departments/programs of Jewish Studies, in departments of Religious Studies, or in Jewish denominational seminaries.[11]

---

[11] See the comments in Aaron W. Hughes and Elliot R. Wolfson, "Introduction: Charting an Alternative Course for the Study of Jewish Philosophy," in *New Directions in Jewish Philosophy*, ed. Aaron W. Hughes and Elliot R. Wolfson (Bloomington, IN: Indiana University Press, 2010), 1–16.

The purpose of the Library of Contemporary Jewish Philosophers is not only to dispel misperceptions about Jewish philosophy but also to help nudge the practice of Jewish philosophy out of the ethereal heights of academe to the more practical concerns of living Jewish communities. To the public at large this project documents the diversity, creativity, and richness of Jewish philosophical and intellectual activity during the second half of the twentieth century, and early twenty-first century, showing how Jewish thinkers have engaged new topics, themes, and methodologies and raised new philosophical questions. Indeed, Jewish philosophers have been intimately engaged in trying to understand and interpret the momentous changes of the twentieth century for Jews. These have included the Holocaust, the renewal of Jewish political sovereignty, secularism, postmodernism, feminism, and environmentalism. As a result, the Library of Contemporary Jewish Philosophers intentionally defines the scope of Jewish philosophy very broadly so as to engage and include theology, political theory, literary theory, intellectual history, ethics, and feminist theory, among other discourses. We believe that the overly stringent definition of "philosophy" has impoverished the practice of Jewish philosophy, obscuring the creativity and breadth of contemporary Jewish reflections. An accurate and forward looking view of Jewish philosophy must be inclusive.

To practitioners of Jewish philosophy this project claims that Jewish philosophical activity cannot and should not remain limited to professional academic pursuits. Rather, Jewish philosophy must be engaged in life as lived in the present by both Jews and non-Jews. Jews are no longer a people apart, instead they are part of the world and they live in this world through conversation with other civilizations and cultures. Jewish philosophy speaks to Jews and to non-Jews, encouraging them to reflect on problems and take a stand on a myriad of issues of grave importance. Jewish philosophy, in other words, is not only alive and well today, it is also of the utmost relevance to Jews and non-Jews.

The Library of Contemporary Jewish Philosophers is simultaneously a *documentary* and an *educational* project. As a documentary project, it intends to shape the legacy of outstanding thinkers for posterity, identifying their major philosophical ideas and making available their seminal essays, many of which are not easily accessible. A crucial aspect of this is the interview with the philosophers that functions, in many ways, as an oral history. The interview provides very personal comments by each philosopher as he or she reflects about a range of issues that have engaged them over the years. In this regard the Library of Contemporary Jewish

Philosophers simultaneously records Jewish philosophical activity and demonstrates its creativity both as a constructive discourse as well as an academic field.

As an educational project, the Library of Contemporary Jewish Philosophers is intended to stimulate discussion, reflection, and debate about the meaning of Jewish existence at the dawn of the twenty-first century. The individual volumes and the entire set are intended to be used in a variety of educational settings: college-level courses, programs for adult Jewish learning, rabbinic training, and interreligious dialogues. By engaging or confronting the ideas of these philosophers, we hope that Jews and non-Jews alike will be encouraged to ponder the past, present, and future of Jewish philosophy, reflect on the challenges to and complexities of Jewish existence, and articulate Jewish philosophical responses to these challenges. We hope that, taken as individual volumes and as a collection, the Library of Contemporary Jewish Philosophers will inspire readers to ask philosophical, theological, ethical, and scientific questions that will enrich Jewish intellectual life for the remainder of the twenty-first century.

All of the volumes in the Library of Contemporary Jewish Philosophers have the same structure: an intellectual profile of the thinker, several seminal essays by the featured philosopher, an interview with him or her, and a select bibliography, which includes such items as books, articles, book chapters, and public addresses. As editors of the series we hope that the structure will encourage the reader to engage the volume through reflection, discussion, debate, and dialogue. As the love of wisdom, philosophy is inherently Jewish. Philosophy invites questions, cherishes debate and controversy, and ponders the meaning of life, especially Jewish life. We hope that the Library of Contemporary Jewish Philosophers will stimulate thinking and debate because it is our hope that the more Jews philosophize, the more they will make Judaism deeper, durable, and long-lasting. Finally, we invite readers to engage the thinkers featured in these volumes, to challenge and dispute them, so that Judaism will become ever stronger for future generations.

MENACHEM FISCH: AN INTELLECTUAL PORTRAIT

Noah J. Efron

Menachem Fisch is the Joseph and Ceil Mazer Professor of History and Philosophy of Science, Director of the Center for Religious and Interreligious Studies, and former Chair of the Graduate School of Philosophy at Tel Aviv University. He is also Senior Fellow of the Kogod Center for the Renewal of Jewish Thought at the Shalom Hartman Institute in Jerusalem. Fisch, scion of an observant Jewish family that integrated Orthodox Judaism, modern Western culture, and Zionism, began his academic career in physics, but was drawn in time to philosophy, and then to history and philosophy of science. From this focus on science, and in particular Victorian science, Fisch confronted questions of broad, almost universal, scope. He inquired into the nature of rational action. He investigated how knowledge is produced. He researched the circumstances that lead people to change their minds and to relinquish cherished theories and worldviews in favor of others. He studied how discourse is possible between people operating in different theoretical or normative frameworks. The answers that Fisch found to these questions proved to be of great moment when he turned his attention to Jewish philosophy. Menachem Fisch has expanded the scope of Jewish philosophy by integrating it into the larger discourse of religion and science, and in turn applied his philosophical, historical, and critical insights to the analysis of Talmudic texts, positing a new and original understanding of rabbinic religious commitment and legal reasoning. As an astute and subtle reader of rabbinic, philosophical, and scientific texts, Menachem Fisch's life project has been to revive traditional Jewish exegetical resources and fashion philosophical resources to free us from ostensible dichotomies that today hobble and bedevil us. He has shown that one need not choose between Jewish particularism and philosophic universalism, between religious observance and philosophic critique, between Zionism and openness to the world. While holding senior academic positions at the Tel Aviv University and advancing the history and philosophy of science, Fisch has not limited himself to academic scholarship but has employed his philosophic acumen to affect change within modern Orthodoxy and between Judaism and other religious traditions, especially Christianity and Islam. As a philosopher of science, intellectual historian, a Talmud scholar-teacher, a constructive theologian, and public intellectual, Menachem Fisch has expanded the scope of Jewish philosophy and deepened our understanding of Judaism. His deep reflections offer a radical political vision that enables contemporary Jews to address profound challenges in Israel, a country riven by internal conflict about its own identity and its attitude toward non-Jews. His vision exemplifies the audacity of humility and the relevance of philosophy to contemporary Jewish politics.

Three Generations of the Fisch Family

Menachem Fisch is the son of a scholar who was himself the son of a scholar. Each Dr. Fisch in each generation was occupied, among his other concerns, with the question of tradition, and how it might be honored and how it might be altered; when its claims on our beliefs and behaviors were to be respected, and when they were to be resisted. Each Dr. Fisch was raised in an observant household and, in the fullness of time, raised his own children in an observant household. While conventional in their Jewish practice, each Dr. Fisch developed unconventional ideas about Jewish tradition and about how Jews and Judaism might meet the modern world. The three generations of overlapping careers have already spanned a century (and Menachem Fisch is only now approaching the height of his powers); the century that saw the final collapse of an old order of Jewish semi-autonomous settlements (with their own languages—Yiddish, Ladino, etc.), and the paroxysms that followed: displacement, persecution, and resettlement of unprecedented scale. Throughout this period, many have pondered, worried about, and sought solutions to what was conventionally called "the Jewish problem," a sobriquet applied to the raft of difficulties faced by Jews trying to adjust to life in the modern West. Rabbi Dr. Solomon Fisch, Prof. Harold Fisch, and Prof. Menachem Fisch were alive to these difficulties; each in his own way devoted his life to addressing them.

Rabbi Dr. Solomon Fisch traversed three worlds.[1] In 1889, he was born into the first, one of modest circumstances and devout Orthodoxy, in Wolbrom, Poland. His father had a small grocery, and when that burned down, he traded in metals. The population of the town was nearly entirely

[1] So wrote Harold Fisch, in the introduction to his father's self-published collection of memoirs and occasional pieces, *Yeriot Shlomo: Pirkei Hayim, Divrei Hagut, Drashot ve-Hiddushei Torah* (Jerusalem, 1983), 12. Indeed, when he published the book, Solomon Fisch was retired in Jerusalem, a fourth "world." However, in the fullness of his days, his new home did not leave a legible imprint upon his thought and work.

Jewish and, Solomon Fisch would later report, he had almost no contact with anyone who was not. His neighbors were peddlers, merchants, and tradesmen.[2] His education was severe and traditional. (This was, Harold Fisch later wrote, "a world of Torah."[3]) In 1913, before the outbreak of World War I, Solomon Fisch resettled in Frankfurt am Main, where he studied with the renowned rabbi Solomon Breuer. Breuer was son-in-law of Rabbi Dr. Samson Raphael Hirsch, whom he succeeded as rabbi of the *Religionsgesellschaft* of Frankfurt, and as the leading light of Hirsch's *Torah im Derekh Eretz* movement (literally "Torah with the Way of the Land"), an effort to render Orthodoxy more consonant with modernity. At the same time, Breuer was known in his day as "the outstanding champion of Orthodoxy in Germany in its fight against the Reform movement."[4] Solomon Fisch embraced Breuer as his mentor, and revered him as uncompromising in his commitment to halakha, yet enlightened about secular matters.[5] At the same time, Solomon Fisch loathed the identification with German culture that he observed among advocates of *Torah im Derekh Eretz*, disparaging in his memoir an Orthodox Rabbi who told him that "in language and dress, we are German, our children at school sit on the same bench as gentiles; only in the synagogue are we Jews."[6]

Two years after the war, in 1920, Solomon Fisch emigrated to Liverpool, where he married. Two years later, he moved to Birmingham, and, four years after that, to Sheffield. It was in England that Solomon Fisch first found employment as a congregational rabbi, a role he fulfilled, with some ambivalence, for decades. To his dismay, he found the Jews in provincial England to be unlearned and, often, only weakly committed to Jewish observance. Wealth, he observed, was often more highly regarded than learning, and community leaders—rabbis and lay leaders alike—were not above petty corruptions. Over the course of his long career, he clashed several times with colleagues who, by his lights, compromised their responsibility and their morality in accommodation of lax local standards. Indeed, the entire structure of the English rabbinate seemed like a diminished

---

2 Me'ir Shim'on Geshuri, *Ṿolbrom 'irenu: sefer-yizḳor li-shenat ha-'eśrim le-ḥurban ha-ayom shel ha-'ir she-hayetah moladetenu ba-'avar* (Tel Aviv: Irgun Yots'e Ṿolbrom be-Yi'sra'el, 2008), 412.

3 Harold Fisch, "Introduction," in *Yeriot Shlomo*, 12.

4 "Rabbi Solomon Breuer, Leader of German Orthodoxy, Dies," Jewish Telegraphic Agency, http://www.jta.org/1926/07/21/archive/rabbi-solomon-breuer-leader-of-german-orthodoxy-dies (accessed October 9, 2015).

5 Harold Fisch, "Introduction," 13.

6 Solomon Fisch, *Yeriot Shlomo*, 38.

appropriation: "the Chief Rabbinate is a foreign plant in the vineyard of the House of Israel and modeled on the Anglican Church," Solomon Fisch wrote.[7] Solace he derived from scholarship. Although he did not hold a bachelor's degree, the University of Manchester admitted him to graduate studies, and he completed a master's degree and doctorate with a critical edition and commentary on the thirteenth-century *Midrash Haggadol*, recovered from a Yemenite manuscript.[8] Along the way, he learned Greek and Arabic.

Solomon Fisch had experienced the ease with which Jews could move from an isolated Orthodoxy to a separatist Orthodoxy to an Orthodoxy integrated into its Christian surroundings; he also experienced the hardships that attended these moves. Secular books shared with religious ones the bookshelves in Solomon Fisch's modest home. The language of the dinner table was English. He addressed area churches about the contributions Jews had made to Christian civilization, and about religious tolerance.[9] In all these ways, his acculturation to England was profound. Still, Solomon Fisch was withering in his criticisms of Reform and Liberal Jews who, no longer saw themselves as obligated by Torah and mitzvot, and for whom the price of entering into Christian society was abandoning all that was unique and valuable in Judaism. These Jews were "worse than Karaites," Solomon Fisch wrote, and "it is forbidden to associate with them."[10] At least until he retired, Solomon Fisch remained an example both of the remarkable integration achieved by some Jews into Western society in the first half of the twentieth century and of the discontents of this integration.

The ease and discontents of modernity presented themselves differently to Harold Fisch, Solomon's son and Menachem's father. Harold Fisch was born in Birmingham, attended a working-class municipal high school, and studied at Sheffield with the revered Shakespeare scholar Geoffrey Bullough. He served as an officer in the Royal Navy at the height of World War II. Pursuing graduate study at Oxford, Harold Fisch was drawn to seventeenth-century divinity, especially as expressed in Anglican poetry:

---

7 Solomon Fisch, "Letter to the Editor," *Jewish Chronicle*, June 26, 1964, p. 7.

8 "Educational Successes," *Jewish Chronicle*, July 3, 1936, p. 37.

9 See, for example, the notice in the *Jewish Chronicle*, April 1, 1938, p. 58: "Rabbi Dr. S. Fisch addressed the Wesley Guild of Methodist Church, in Woodhouse, on 'Jew and Gentile.' The Rev. E. J. Prentice, who presided, referred sympathetically to the plight of Austrian Jewry. Among those who took part in the discussion were the Rev. S. Snell, Vicar of Woodhouse, and Councilor C. G. Mitchell."

10 Solomon Fisch, "'Am Segulah," in *Yeriot Shelomo*, 88.

"It may be wondered how this participation in the spiritual world of a seventeenth-century Anglican divine chimed with my Jewishness?"[11] Years later, Harold Fisch reflected on the relationship between the literature he studied and the religious he practiced:

> Was I not in danger of crossing boundaries and compromising my heritage of Jewish beliefs and practices? The answer is: quite the contrary! This deepening interest in seventeenth-century English divinity proceeded simultaneously with a growing Jewish awareness ... The two processes converged ... What I was discovering in the early seventeenth-century poets and preachers was a reflecting glass that made my Judaism clearer to me ... I was ... constantly discovering matters that belonged to my own religious tradition, but presented in an unfamiliar, and hence more interesting guise.[12]

There were complexities in Harold Fisch's understanding of the relationships between Western, Christian traditions and the Jewish traditions that they mirrored, and there were tensions and fault lines that remained throughout his career. After Oxford, in 1947, Harold Fisch took an appointment at the University of Leeds, where his father had accepted a pulpit a year earlier. There, he writes, "I was able at this time to benefit from my father's scholarship."[13] In particular, Harold Fisch embraced Solomon Fisch's "conviction of the absolute moral value that attached itself to the notion of 'the Jewish People.'"[14]

> Jewish peoplehood was a fundamental condition of our existence, a source of identity both for the individual and the collective .... Whether we were in Poland or Germany or England, we were a people in exile. To imagine ourselves as belonging to one of these other nations was to court ruin and disaster.[15]

This conviction of the "absolute moral value" of Jewish peoplehood was behind Harold Fisch's 1957 decision to accept a professorship in English at Bar-Ilan University and to move with his wife and four young children to an Israel that had not yet celebrated a decade of independence. It was also a force in his decision a decade later, immediately after the 1967 Six-Day War, to participate in the establishment of "The Land of Israel Movement," which

---

11 Harold Fisch, *Who Knows One?: An Essay in Autobiography* (Ramat Gan: Bar-Ilan University Press, 2004), 45. He wrote his B.Lit. thesis at Oxford on the Calvinist Bishop Joseph Hall (1574–1656).

12 Ibid., 45–47.

13 Ibid., 54.

14 Ibid.

15 Ibid., 54–55.

aimed to ensure that the territories occupied in that war "would remain the inalienable possession of the Jewish People."[16] Harold Fisch's delight in the fact that the works of Milton, Shakespeare, Blake, and other luminaries of English literature illuminate, and can be illuminated by, canonical Jewish texts, sat uneasily with his certitude that the Jewish people were and remain sui generis. There was a "great divide between the outward openness of his scholarly enterprises and the inward self-sufficiency of his Judaism," as Menachem Fisch would later write. This "was a tension that was never resolved,"[17] although it was addressed, in the final weeks of Harold Fisch's life.

Harold Fisch's final public appearance came at a 2001 academic gathering in his honor. There, the elder Fisch spoke of the biblical tale of Jacob wrestling with the angel, a story that ends with the patriarch being assigned the additional name, Israel ("he who prevails with God"). After observing that Jacob is the name accorded to the patriarch at his meekest, and Israel is the name accorded to the patriarch at his most combative and triumphant, Harold Fisch mused about what lesson this nominal dialectic offered for his own day. There are those, he observed, who believe that Jews must embrace the combative legacy of Israel first, diminishing the soft conciliation of Jacob. Others believe the opposite. "There is no doubt in my mind," he continued, that Israel must be the principal name of Jews in our time, and Jacob of secondary importance. However, others of wisdom and discrimination arrive at the opposite conclusion. It was at this moment, in his last public address, that Harold Fisch asserted, as his son later described it, that "we are required to confront and be willing to contend with the dialectical fullness of the tradition that does *not* speak in one voice."[18] Harold Fisch spoke and wrote with refinement, his literary criticism was admired throughout the world, and over his long and storied career, he found appreciative audiences in the finest universities in the West. Still, there was an incommensurability to the different worlds that he inhabited, a tension that was never fully resolved.

Biography and Career

Menachem Fisch was raised with this tension. He was almost nine when his father accepted his position at Bar-Ilan University, and the family moved

16 Ibid., 130.
17 Menachem Fisch, "Introduction," in *Who Knows One?*, 22.
18 Ibid., 20.

from Leeds to Petach Tikvah. In time, he attended the Midrashiyat Noam, a school established in 1945 as an offshoot of the No'ar Mizrahi youth movement. The school, like the movement, embraced the principles of *Torah im Derekh Eretz* and aimed to demonstrate the harmony between a life of traditional Jewish learning and practice and a life of engagement in the secular world. Such harmony as existed, however, was superficial and of a mechanical sort. The school devoted mornings to traditional study of classic Jewish texts and afternoons to secular subjects, but little effort was made to link the morning's studies to those of the afternoon, or the reverse. At best, Menachem Fisch remembered, "secular subjects were implicitly taught as merely instrumental to getting ahead in society." Among secular subjects, "the sciences were considered relatively safe for Orthodox Jews, while the humanities were perceived as posing a threat to religion."[19] Upon completing secondary school, Menachem Fisch enrolled to study physics and mathematics at Bar-Ilan. Upon completing his undergraduate studies, he entered a graduate program in physics, and had almost completed a Master's Degree when he realized that he did not wish to become a physicist.[20]

Leaving Bar-Ilan, Menachem Fisch was inducted into the IDF Artillery Unit (having earlier deferred service for studies). After three years of mandatory service, which included officers' training, he was pressed by his commanders to extend his service. Three years later he retired from active service as a captain, rising swiftly to lieutenant colonel in the reserves. Fisch determined to pursue advanced studies in philosophy of science, matriculating at Tel Aviv University. In time, he produced a master's thesis on the "Ravens Paradox," a conundrum proposed by Carl Hempel a generation earlier, raising questions about what does and does not count as evidence for generalizations based on observations; these questions had occupied analytic philosophers of science ever since. Fisch's treatment was original and sound, and it was later published in the form of two well-regarded peer-reviewed articles.[21] It also persuaded its author that there were limits to the reach of analytical philosophy of science. "Philosophy of science," Fisch concluded, "should not be pursued in the abstract,

---

19 See the interview in this volume, p. 157.

20 Fisch's predilection for philosophy had already been noted in an article about student life that appeared in the Labor movement newspaper *Davar*, which took pleasure in the fact that the physics student from a religious university kept volumes by Jean-Paul Sartre at his bedside. Michal Hamburg, "Ian Fleming and Jean-Paul Sartre," *Davar*, April 15, 1969, p. 4.

21 Menachem Fisch, "Hempel's Ravens, The Natural Classification of Hypotheses and the Growth of Knowledge," *Erkenntnis* 21 (1984): 45–62, and Menachem Fisch, "Whewell's Consilience of Inductions—An Evaluation," *Philosophy of Science* 52 (1985): 239–55.

but in a way intimately connected to science as it is and was lived and done, combined much more closely with the history of science."[22] This was a realization that set him off on a new path.

Fisch's doctoral dissertation was an effort to understand the life's work—the *project*—of William Whewell, a great nineteenth-century English polymath who made important contributions to philosophy (and especially philosophy of science), history of science, theology, and more. The early 1980s, when Fisch carried out his research (at Tel Aviv University and The Queen's College, Oxford, where he was a visiting graduate student), was a time of enthusiasm among historians, philosophers, and sociologists of science for describing the intellectual achievements of past savants as products of the social, cultural, and ideological environments in which they were accomplished. While attentive to the power that such forces exert on human thought, Fisch took a different approach. He set out to illuminate Whewell's work by reconstructing the *process of deliberation* that guided him. Fisch sought to describe the "erotetic logic" of Whewell's inquiries, namely, the chain of questions he posed to himself, the answers he considered, and the new questions and challenges these raised. He accomplished this painstakingly, through reviewing successive drafts of Whewell's works, taking careful account of the changes he affected, and reconstructing the questions these changes answered and the problems they solved. In so doing, Fisch replaced a generally static picture of Whewell with a dynamic one, of a scholar struggling forth, seeking haltingly to understand his subjects of study, putting forth hypotheses only to abandon some and refine or reframe others. Whewell's import, Fisch observed, lay mostly "in the process by which his views took shape," rather than in the impact of his views themselves.[23] In Fisch's presentation, Whewell was a man of his times, but he was at the same time a man endeavoring to use reason to answer questions and resolve problems. Eventually published as *William Whewell: Philosopher of Science*, Fisch's research defied easy characterization.[24] It was neither "externalist" nor "internalist," the names given to the two main camps of historians of science, the one which placed its emphasis in the context in which science was pursued, and the other on the succession of ideas that led a new scientific theory to displace its

[22] See the interview in this volume, p. 162.

[23] Menachem Fisch, *William Whewell: Philosopher of Science* (Oxford: Clarendon, 1991), 8.

[24] Fisch, *William Whewell: Philosopher of Science*. See also Menachem Fisch and Simon Schaffer, eds., *William Whewell: A Composite Portrait* (Oxford: Clarendon Press, 1991).

predecessor. It was, instead, a study of how a person labors to make sense of the world around him, to absorb the insights of others into his own thought and, on occasion, to challenge and change conceptions that he adopted or created in the past.

In the following years, Menachem Fisch continued his research on Victorian natural philosophers and mathematicians, principally Whewell, John Herschel, Charles Babbage, George Peacock, and James Clerk Maxwell. At the same time, he deepened his philosophical study of rationality and rational action. The influence of Karl Popper on this work is evident. Fisch embraced and expanded Popper's "critical rationalism," the assertion that our confidence about the truth of a proposition grows not by adducing supporting evidence, but by subjecting it to the staunchest criticism we can muster. Popper argued that it is by withstanding vigorous attempts at *falsification*, rather than being bolstered by confirmation, that a proposition becomes more worthy of acceptance. These investigations into rational activity brought together concerns that had occupied Fisch for the first decade of his academic life, in his master's thesis on Hempel's paradox of induction, in his doctoral dissertation on Whewell's intellectual trajectory, and in his work on Victorian science more generally. Fisch's persuasion that criticism was at the heart of rational inquiry became increasingly central to his graduate teaching at this time, in two ways. Growing portions of his syllabi were devoted to exploring the nature of criticism and, at the same time, criticism came to animate his pedagogy. His advanced graduate seminars were devoted to close, critical readings of historical scientific and philosophical texts, in which students (of whom, by this point, I was one) were asked to subject canonical interpretation to criticism, identifying problems in accepted readings, producing alternative readings, and then subjecting them to further scrutiny. The surest way to impress Fisch was to find fault, compellingly, with his own interpretation of a text, or to offer an alternative interpretation.

Fisch's work on history and philosophy of science was recognized by Tel Aviv University, his academic home since 1985. On the faculty of Tel Aviv University, Fisch became a member of the Cohn Institute for the History and Philosophy and Science and Ideas, a research and graduate teaching institute within the Faculty of the Humanities. Established by the late Professors Yehuda Elkana and Amos Funkenstein, and bolstered in 1989 by support from Barbara and Bertram J. Cohn, this academic unit has focused on history of science, intellectual history of the Western world, philosophy of science after the scientific revolution, anthropology of knowledge, history and philosophy of biology, history of technology, history and philosophy of ancient and modern mathematics, cultural studies, and critical study of science, society, and religion, awarding doctorate degrees to over one hundred students. Fisch holds the first doctorate awarded by the Cohn Institute and is currently the Chair of the Cohn Institute's Board and Academic Committee. In this capacity he has enhanced the international reputation of the Cohn Institute through collaboration and affiliation with similar academic institutes worldwide. Over the years Fisch has held visiting research positions at The Queen's College, Oxford; Trinity College, Cambridge; the Dibner Institute for Advanced Studies in the History of Science and Technology at MIT; the Institute for Advanced Study in Princeton; Collegium Budapest; and the Frankfurt University Institute for Advanced Studies in the Humanities. He has contributed to the field of history and philosophy of science through editorial work as co-editor of *HOPOS: The Journal of Internal Society for the History of the Philosophy of Science.* His international reputation was recognized in Israel when he was appointed in 1999 founding President of the Israel Society for History and Philosophy of Science, a position he held until 2004. He concurrently served as Chair of the Israeli National Committee for History of Science.

Reading Rabbinic Judaism Philosophically

Fisch's outstanding contribution to the field of history and philosophy of science has never negated his religious observance or his commitment to the legacy of *Torah im Derekh Eretz,* which he inherited from his father and grandfather. In the late 1980s he began to apply his philosophical training to Jewish texts. In November 1989, Bar-Ilan University organized a conference honoring the work of Harold Fisch, and Menachem Fisch was pressed to present a paper in tribute to his father. His tribute was a challenge. The elder Fisch had, not long before, published an essay about Ecclesiastes entitled, "Qohelet: A Hebrew Ironist." In it, he argued that the book of Ecclesiastes presented a rebuke of those who eschewed God and relied instead on human rationality and wisdom alone. In Harold Fisch's reading, the book of Ecclesiastes argued that this was a futile vanity (ending with this charge: "The end of the matter, all having been heard: fear God, and keep His commandments; for this is the whole man." [12:13]). With his father seated in the first row, Menachem Fisch challenged this interpretation, asserting instead that the book could be read (and indeed had been read by the rabbis who canonized it) as arguing not that knowledge attained through human effort alone is futile, but that it is tentative and uncertain. A linchpin of his alternative interpretation was the word *hevel.* Traditionally, this was taken to mean "foolishness" or "futility." However, the younger Fisch pointed out, the word's original meaning was "misty" or "vaporous." This being the case, Ecclesiastes' brief was not against seeking wisdom through rational means, but rather against dogmatism, and the belief that absolute knowledge had been attained or is attainable in principle. This, Menachem Fisch has said, was the "moment that launched me in the direction of thinking seriously about Judaism through the prism of Talmudic reasoning."[25]

In 1995, Fisch published a slim Hebrew volume entitled *To Know Wisdom: Science, Rationality, and Torah Study.* The book was based on a series of acclaimed public lectures that Fisch had given at the Jerusalem Van Leer Institute, and argued that Talmudic literature, at least from the second generation of *Tana'im,* sought to document traditions of learning founded upon notions of rational and progressive inquiry that were very similar to notions at the foundation of modern scientific practice. The protocols of rabbinic discussions contained in the Babylonian Talmud, Fisch argued, demonstrate the rabbis' radical willingness to subject truth claims to vigorous criticism. Further, Fisch argued, the Book of Ecclesiastes, read as he proposed, explicated the epistemological commitments of the rabbis that were on display in the Babylonian Talmud; it served them as a sort of handbook about how best to seek knowledge, and they applied its lessons in the discussions compiled in the Talmud. The epistemology that Ecclesiastes described and the Talmud reflected was, broadly, the one described (albeit crassly) by Karl Popper: an epistemology of constructive skepticism and rational action driven by criticism.

Even before *To Know Wisdom* reached bookstore shelves, Fisch was at work on a new book that set out his arguments in greater detail and with greater rigor: *Rational Rabbis: Science and Talmudic Culture.* (A précis of the book is reprinted in the present volume: "Rational Rabbis: Its Project and Argument.") The new book contained two unequal sections. The first set out a general theory of rational endeavor. Popper's epistemology of constructive skepticism served as a point of departure, but the argument made clear how far Fisch had by now moved beyond his erstwhile inspiration. For one thing, at the foundations of Popper's accounts of

---

25 See the interview in this volume, p. 165.

science were concepts that Popper himself had never unpacked: "criticism" and "rationality," most obviously, and also "problem" and "solution." In the first part of *Rational Rabbis*, Fisch provided a more comprehensive account of rational endeavor than Popper ever had. Such endeavor, he wrote, takes place within the context of a "goal-directed system," meaning "any structure, theoretical, institutional or material, designed or adopted by human agents as a means to specific ends." A "problem" is the failure of such a system to reach its goals, a gap between what a system is meant to do, and what it, in fact, accomplishes. One subjects a goal-directed system to criticism by seeking to identify and understand its problems. When one has identified and understood problems in a system, they can be fixed (or, at the very least, one can try to fix them). Fixing problems is the way that one improves a system. This improvement can be seen as progress. And, Fisch writes, "to act rationally... is to contrive deliberately to effect progress."[26]

This simple, lucid account of rational action proved to be remarkably powerful. It gave good account of the behaviors of scientists in laboratories and equally explained the behaviors of a chef tinkering with a recipe, a novelist revising a chapter, and a coder debugging a program. In the second part of *Rational Rabbis*, Fisch argued that this account of rational endeavor elucidates as well the discourses and decisions of Talmudic sages. He identified in Talmudic discourse two camps, one "traditionalist" and the other "antitraditionalist." For those in the first camp, once established, a halakhic tradition is unassailable and unalterable. For those in the second camp, halakhic tradition must be constantly scrutinized and, if it proves inadequate or faulty, revised and improved. This latter, antitraditionalist approach, Fisch observed, is "a position closely akin to the neo-Popperian view of rational endeavor" that he earlier described. In a series of close reading of Talmudic *sugyot* (discursive passages), Fisch identified time and again this "clear and central" antitraditionalist voice. *Rational Rabbis* was sometimes misinterpreted as an argument that there is a deep affinity and, hence, compatibility between scientific and Talmudic reasoning. This misinterpretation appealed to some, who wished to explain the distinguished accomplishments of Jews in modern sciences by attributing them to a Talmudic tradition that somehow primed them to excel at science.

---

[26] Menachem Fisch, "Rational Rabbis: Its Project and Argument," *Journal of Textual Reasoning* 4, no. 2 (March 2006), http://jtr.lib.virginia.edu/volume-4-number-2/rational-rabbis-its-project-and-argument/.

In fact, Fisch offered no support for such apologetics. His claim was that both scientific and Talmudic discourses are the products—in part, at the very least—of rational human endeavor. It was in this way that, as one scholar put it, "*Rational Rabbis* announces the collapse of yet another wall separating talmud and talmudic study from the minds of modern western intellectuals."[27]

In *Rational Rabbis*, Fisch had also uncovered a deeper issue that he would explicate in the fullness of time. Not only does the Talmud include two very different sorts of voices, each born of and bearing a different theory of how knowledge is derived and how durable it can be known to be, but it suggests that these two voices bear an important relationship to one another. Rereading *Rational Rabbis* almost two decades after it was first published, especially in light of Fisch's work in the intervening years, one identifies a claim, not yet fully articulated, that the complicated relationship between the traditionalist and antitraditionalist perspectives Fisch describes is more dialectical than antagonistic. Traditionalist and antitraditionalist positions are juxtaposed. The continuing conflict between them is presented itself as an exemplar of dialogical and critical reasoning. But what's more, the positions are somehow mutually constitutive—each helps to define the other. This is the seed of an idea that would flower fully in Fisch's later work.

Fisch's original and creative readings of rabbinic texts have been scrutinized, tested, disputed, and enriched by thousands of hours of discussion, learning, and teaching at the Shalom Hartman Institute in Jerusalem, where he has been Senior Research Fellow since 1993. The Shalom Hartman Institute was founded by Rabbi David Hartman in 1971 to address signal challenges of the day: to foster discussion among people of different beliefs; reenvision Jewish sovereignty and political power in ways that draw upon, and expand, traditional Jewish intellectual and spiritual resources; devise new models of Jewish learning; integrate women fully into Jewish practice and thought; and establish new modes of Jewish learning. The Institute describes itself as "a center of transformative thinking and teaching." Its fellows comprise a cadre of scholars who, at the Institute and at their home universities across the country, have done much to redefine Jewish philosophy in our day. (Aside from Menachem Fisch's work,

[27] Jacob Meskin, "Reasonable Rabbis?! Practice and Situatedness in Menachem Fisch's *Rational Rabbis*," *Journal of Textual Reasoning* 4, no. 2 (March 2006), http://jtr.lib.virginia.edu/volume-4-number-2/reasonable-rabbis-practice-and-situatedness-in-menachem-fischs-rational-rabbis-jacob-meskin-hebrew-college/.

the work of two other Hartman Institute fellows has been celebrated in volumes of the Library of Contemporary Jewish Philosophers: Moshe Idel and Avi Sagi.) The Hartman Institute has been a laboratory for Fisch, where he has tested and refined ideas with rabbis, teachers, lay leaders, scholars and clergy from other religions, and some of the greatest Judaic scholars of the age. There he edits, with Moshe Idel, a book imprint called *Yahaduyot* (Judaisms) and organizes an annual conference on theology, which is described below.

Science and Religion

By the time Fisch completed *To Know Wisdom* and *Rational Rabbis*, his scholarly interests in philosophy, intellectual history, Judaics, and religion more generally had fused in a reaction that continues to emit great energy to this day. In an essay of reach and ambition called "Judaism and the Religious Crisis of Modern Science" (included in this volume), Fisch describes the "now familiar tensions between science and religion" as a result of an historical process that began with the reformation. Intellectual historian Peter Harrison had famously argued that early modern natural philosophy (what would one day come to be called science) owed much to the insistence of Luther and other Protestant reformers that the truth of the Bible was reflected in a literal reading of the book which could be accomplished by any prudent reader. The way the reformers insisted Bible be read became, mutatis mutandis, the way natural philosophers came to insist that nature be read. Seeing a link between Holy Writ and creation was not new. Christian theologians had since Augustine been accustomed to viewing Bible and nature as parallel subjects of inquiry; the "Book of Scripture" and the "Book of Nature" were the "two books" authored by God and through which God revealed himself. Harrison's argument was that once Protestants began to read in new ways the "Book of Scripture," they came soon to read in similarly new ways the "Book of Nature." Just as the Protestant Bible reader stripped Holy Writ of its symbolic, metaphoric, and allegorical meanings (settling instead for straightforward literality), so too the Protestant reader of nature stripped the subject of his inquiry of symbolic, metaphoric, and allegorical meanings (settling instead for straightforward factuality). To Fisch, this explanation was illuminating, but also partial, and he set out to expand it. It is true, Fisch wrote, that a Protestant account of science, best exemplified by the philosophy of Francis Bacon, "insisted on grounding scientific truth in the immediacy of factual experience from which one then reasons *up* inductively to theory." However, there was also, in the seventeenth century, a Catholic Counter-Reformation account of science, best exemplified by Descartes, which "insisted, equally vigorously, that it be grounded in the *a priori* certitude of the mathematicized systems of ideas given by pure reason, from which one then reasons *down* deductively to the facts."[28] In each approach, the account of science remained congruent with the religion of its advocate.

This state of affairs changed, Fisch writes, after Newton. Newton's magisterial *Philosophiæ Naturalis Principia Mathematica* (1687) combined elements of Bacon's Protestant approach and of Descartes' Catholic approach without comment and without trying to harmonize the seemingly opposed views. As Fisch put it, the great book "exhibited a perplexing mixture of mutually irreducible empiricist and rationalist elements that neither philosophy of science was capable of accommodating from its own purist perspective."[29] As a result, "in the heady wake of Newton's great achievement, philosophers found themselves having to abandon major elements of the two philosophies of science that had launched modern science on its way."[30] When these philosophies of science lost their purchase, so too did other matters of consequence. The link was snapped between the way Bible was best read and the way the world was best studied. With that, the "Two Books" metaphor lost its currency. Science and Christianity (both Protestant and Catholic forms) went from being mutually supportive endeavors to being mutually estranged endeavors. Science moved away from its earlier foundationalist dreams (of being planted in the rich, firm soil of observation, as Bacon hoped, or in the rich, firm soil of first principles derived from reason, as Descartes hoped), at the same time that Christianity moved towards foundationalism (of revelation coded in unchanging dogma). As these tectonic philosophical shifts progressed, Christianity came to seem simply irrational in the eyes of those for whom science served as lodestar. And science came to seem wantonly corrosive of received wisdom, in the eyes of those for whom Christian revelation served as a lodestar. It was in these changes that one finds the roots of

---

[28] Menachem Fisch, "Judaism, and the Religious Crisis of Modern Science," in *Nature and Scripture in the Abrahamic Religions*, vol. 2, ed. Jitse M. van der Meer and Scott Mandelbrote (Leiden: Brill, 2008), 533.

[29] Ibid., 534.

[30] Ibid., 536.

mutual alienation and mutual antagonism that have so often characterized relations between modern science and modern Christianity.

There is audacity in this analysis; it is the product of scholarly mastery over a sprawling corpus of early modern philosophy and theology, combined with an uncanny ability to see the patterns that make up traditions of thought in transition. Like new-fangled technologies that allow us to sharpen focus onto any element in a digital photograph, Fisch's description portrays forest and trees without occluding one for the other. A virtue of the explanation that he provides is that it shifts the attention of those seeking to understand tensions between modern science and modern Christianity from ostensible conflicts between concrete beliefs about the world (say, scientists claiming that the age of the earth is measured in eons, and clergy claiming it is measured in millennia), to disagreement about how reliable knowledge is attained. To this day, many people believe that when scientists and religionists disagree, it is over how the world *is* and how it came to be. Often, though, it is over how we *make sense* of the world. Framing matters more broadly still, Fisch writes that:

> modernity's great challenge to religion has, therefore, far less to do with questions concerning the age of the universe, or with squaring neo-Darwinism with Scripture, than with the accusation that in submitting to their faith's demand for epistemic commitment, religious individuals were surrendering something central and constitutive of their very humanity: the healthy robust self-questioning of the truly rational, that finds paradigmatic expression and warrant in the shining example of modern science.[31]

But modernity's great challenge to religion does not challenge all religions equally. Developing the themes he explored in *Rational Rabbis*, Fisch writes that even as science in Newton's wake grew alien to many Christian savants, it was growing closer to the views and practices of many Jewish savants. To illustrate, he cites a mid-nineteenth-century passage by R. Naftali Tzvi Yehudah Berlin:

> just as it is not possible for the wise student of nature ever to boast knowledge of all of nature's secrets . . ., and just as there is no guarantee that what his investigations do accomplish will not be invalidated in this generation or the next by colleagues who elect to study the same things differently, so it is not possible for the student of the Torah ever to claim that he has attended to each and every point that claims attention, and even that which he does explain—there is never proof that he has ascertained the truth of the Torah![32]

---

[31] Ibid., 539.
[32] Ibid., 541.

This view, Fisch observes, is "commonplace" in Jewish intellectual history: "The idea of there being a single, true and humanly discernable reading of the biblical text is as foreign to the Jewish exegetical tradition as it was central to early modern Christianity."[33] But there is more to a religion than the way it reads Holy texts. When it comes to "the legislative, judicial and performative aspects of Jewish law, … its norms of conduct [and] what Jews are required to do and to refrain from doing"[34] there is no consensus in favor of the epistemic humility espoused by Berlin. In fact, in matters of halakha, even Berlin believed that there is little place for doubt, "On the question of obedience and submission to … halakhic authority," Fisch writes, "the tradition … is deeply and intriguingly divided" between a "submissive" or "traditionalist" approach, on the one hand, and a "confrontational" or "antitraditionalist" approach, on the other.[35] Indeed, the dispute between these two approaches to halakha "is foundational to the entire Talmudic project."[36] In a sense, the tension between these two approaches is one of the things that the Talmud is *about*. "Judaism's foundational project of religious self-understanding produces not one, but two radically different self-readings."[37]

Pluralism and Tolerance: The Audacity of Humility

It is in one of those "self-readings," the "confrontational" or "antitraditionalist" one, that Fisch finds a foundation for a new conception of pluralism and a new conception of tolerance. Tolerance is typically grounded on one of two notions. One is that it is a damnable necessity if people are to live peaceably; this view Fisch calls *Real Politik*. The other is that human autonomy—the freedom to think and believe what we wish, so long as we do not hurt others—is a value of surpassing importance, and respecting the autonomy of others demands that we tolerate their views, no matter how mistaken they seem to us. This view Fisch calls "liberal." In both formulations, tolerance retains much of its original meaning as something that must be borne, in irritation and discomfort ("the action or practice of enduring or sustaining pain or hardship," according to the Oxford English Dictionary). Both notions of tolerance aim to answer the question that

[33] Ibid., 543.
[34] Ibid., 546.
[35] Ibid., 547.
[36] Ibid., 559.
[37] Ibid., 563.

Isaiah Berlin famously identified with what he called "negative liberty": "What is the area within which the subject—a person or group of persons—is or should be left to do or be what he wants to do or be, without interference by other persons?"[38] We *tolerate* what we can neither appreciate nor rid ourselves of.

In two essays reprinted in this volume—"A Modest Proposal: Toward a Religious Politics of Epistemic Humility" and "Judaism and the Religious Value of Diversity and Dialogue: Drafting a Jewish Response to *Nostra Aetate*"—Fisch describes how the "antitraditionalist" voice so evident in the Talmud, and which runs through subsequent rabbinic literature like a golden thread, goes far beyond even the most expansive notions of tolerance. "Antitraditionalists, motivated by perpetual self-doubt, are constantly on the lookout for possible flaws and problems in their own system."[39] The "epistemic humility" displayed by antitraditionalists spurs them to embrace "a healthy, religiously meaningful, critical pluralism that allows one to both keenly disagree with, and, at the same time, seriously value the existence of... opinions opposed to one's own."[40] The *other* is not simply one whose errancy must be allowed for the sake of peace or because they have a right to be mistaken; the other is an asset. The other challenges our assumptions, upsets our complaisance, goads us to cast doubt on accepted wisdoms, and needles us to improve upon our system of beliefs. The reason to make room for views different from one's own need not be *grit-your-teeth-and-bear-it* toleration. The reason can be that confronting other views allows us to improve our own. Perspective from within a religious tradition with which one's relations have, historically, been uneasy, can have positive *religious* value, as they can render more coherent, cohesive, and compelling one's religious views.

Such a view has radical implications. Just over fifty years ago, the Second Vatican Council adopted two documents affirming the commitment of the Catholic Church to tolerance: "Declaration on Human Freedom" and "Declaration on the Relation of the Church to Non-Christian Religions" (or *Nostra Aetate*, "In Our Times"). In the first document, the Church asserted that "freedom has roots in divine revelation" and, as a result, it is wrong to coerce people to believe in a particular way, or to accept this or that doctrine. In the second, the Church denounced

[38] Isaiah Berlin, *Four Essays on Liberty* (London: Oxford University Press, 1969), 24.

[39] Menachem Fisch, "A Modest Proposal: Toward a Religious Politics of Epistemic Humility," *Journal of Human Rights* 2, no. 1 (March 2003): 60.

[40] Ibid., 60–61.

anti-Semitism and repudiated the charge that Jews bore collective guilt for the crucifixion of Jesus. It then went further, urging Catholics to speak in earnest, and work in tandem, with believers of different religions, and to "recognize, preserve and promote the good things, spiritual and moral, as well as the socio-cultural values found among these men."[41] Reflecting on the declaration on its jubilee, Pope Francis observed that, in its wake, "indifference and opposition have turned into collaboration and goodwill: from enemies and strangers, we have become friends and brothers."[42]

In finding value in the spiritual, moral, social, and cultural practices of others, the Vatican documents went beyond a standard liberal account of tolerance. They suggested that Catholics can do better than simply suffer others and their errant ways; they can love these others and see in them traits of universal value. Still, they fall short of the sort of tolerance that might find sanction in rabbinic antitraditionalist thought. For such antitraditionalism

> full-consciously locates the definitive corrective, the ultimate normative check on both Judaism's system of law and a Jew's personal conduct, in the moral critique leveled at them by the civilized other. Self-critique, especially normative self-critique is severely limited, especially with regard to basics. And it is here, in the moral challenge leveled at us by adherents to ways of life very different from, even contrary to our own, that the Talmud locates the ultimate source of moral correction.[43]

In such a system, members of one religious community do not merely endure members of another in order to keep the peace. Neither do they merely celebrate others for their universal virtues and humanity, as Vatican II counsels. In such a system, members of one religious community embrace others because these others *help* in crucial ways. Others raise questions about what we do and how we think, and they present alternative ways to live in the world and to make sense of it. Others are to be sought, because they can teach us things (and unteach us things) that we cannot learn (or unlearn) on our own.

---

[41] Menachem Fisch, "Judaism and the Religious Value of Diversity and Dialogue: Drafting a Jewish Response to *Nostra Aetate*," *Diversität—Differenz—Dialogoizität: Religion in Pluralen Kontexten*, ed. S. Alkier, M. Schneider, and Ch. Wiese (Berlin and Boston: de Gruyter, forthcoming).

[42] Ann Schneible, "'We must pray': Pope Francis Declares in Prayer, We 'ask for the gifts for which humanity yearns,'" *Catholic Online*, Oct. 29, 2015, http://www.catholic.org/news/hf/faith/story.php?id=65065http://www.catholic.org/news/hf/faith/story.php?id=65065.

[43] Fisch, "Judaism and the Religious Value of Diversity."

This matter is "worryingly relevant to the modern Israeli," Fisch writes in "A Modest Proposal." Maimonides had been of two minds about Jewish toleration of gentile "idolaters," suggesting acceptance of others when Jews "live exiled among the nations," but very different attitude when Jews gain sovereignty:

> When Israel is predominant over the nations of the world, we are forbidden to permit a gentile who is an idolater to dwell among us. He must not enter our land, even as a temporary resident; or even as a traveler, journeying with merchandise from place to place, till he has undertaken to keep the seven Noachian Laws. (*Hilkhot Avodat Kokhavim* 10:5–6)[44]

The harsh tenor of Maimonides' judgment is alive in the attitudes of some rabbis and politicians who exert influence on present-day Israeli policy. Fisch quotes a Chief Rabbi of Israel, Eliyahu Bakshi-Doron, who advised travel agents "to refrain from encouraging Christian religious tourists or assisting them in visiting and performing idolatry in the places sacred to them."[45] Such a view, when applied to government policy, of course contravenes elementary principles of liberal, multicultural democracy, which guarantees freedom of worship. But that is not the only reason why it is objectionable. For, as Fisch explains, there are religious reasons to encourage pluralism. Judaism itself does not benefit from expelling or silencing non-Jewish voices in a Jewish state. This is because

> belief systems and lifestyles other than their own are given a voice and allocated space by members of the open society primarily because they realize that a plurality of engaging, especially conflicting voices provides the best setting for most effectively articulating and developing one's own. The close presence of challenging, even disturbing and dissenting voices is welcomed, because the chances of one's awakening to error and oversight are exceedingly greater when faced with a variety of real and serious alternatives.[46]

Fisch's analysis has special resonance in an Israel increasingly divided between secular liberals who ground their calls for pluralism and tolerance in a Rortian relativism, and illiberal believers who seek to delimit pluralism and tolerance as threats to cherished truths. His analysis speaks with equal force to both camps and promotes an attitude towards others that is more respectful and appreciative than either camp is able to muster today.

---

44 Quoted in Fisch, "Modest Proposal," 52.
45 Fisch, "Modest Proposal," 53.
46 Ibid., 55.

Political Theory

Viewed more broadly, Fisch's approach has sweeping implications for normative political theory. Tolerance has been, of course, a matter of ambivalence and paradox probably for as long as people have written about politics. It was debated by classical philosophers and pondered by Church Fathers and their disciples. The implacable traumas of the Reformation, Counter-Reformation and religious wars left seventeenth- and eighteenth-century European intellectuals in dire need for a workable theory of toleration. When Rousseau dryly observed in *On the Social Contract* that "it is impossible to live in peace with those one believes to be damned," the weight of centuries of European history ratified his remark.[47] Only months after the start of the French Revolution, in December 1789, the Assembly met to decide how properly to emancipate the Jews. Count Stanislaw de Clermont-Tonnerre, who advocated emancipation, formulated this famous dictum: "As a nation the Jews must be denied everything, as individuals they must be granted everything."[48] This was an extreme example of a notion that found traction in nineteenth-century Europe and left its imprint across the Atlantic in the United States as well. The radicality of this solution says something about the severity of the problem. For the "Jewish problem," as it was everywhere called, was a real problem indeed. In its narrow form, it asked how Jews, as Jews, could be absorbed into liberal society. In its more general form, it asked how people with values and identities that are at odds with those of any given society could find a place as equals in that society, without corroding it. By what mechanism could a society give an equal break to people who insist that they are somehow different from the rest?

This was a problem addressed with varying degrees of success by generations of liberal political theorists,[49] from Thomas Paine, Thomas Jefferson, and John Adams, through Wilhelm von Humboldt and John Stuart Mill to Isaiah Berlin and Friedrich Hayek. In 1971, when Harvard political theorist John Rawls published *A Theory of Justice*, there were many who concluded that, with this book, the problem was pretty much solved. Rawls wrote that a just society is a fair society, and that what's

---

47 Jean-Jacques Rousseau, *On the Social Contract* (New York: Hackett Publishing, 1988), 102.

48 Quoted in David Vital, *A People Apart: A Political History of the Jews in Europe 1789–1939* (Oxford: Oxford University Press, 1999), 42.

49 As well, of course, as nonliberal political theorists, principal among them Karl Marx, whose approach to this question (sadly, some will say) finds little purchase in mainstream contemporary political thought.

fair is to embrace principles and rules that everyone would agree to *before* they knew whether these principles and rules would selfishly benefit them. Under these conditions, people would choose to safeguard their freedoms. Under these conditions, people would fashion a liberal society with a heart, one that might allow inequality, but would insist that the unfortunate benefit from this inequality alongside the fortunate.[50]

Rawls's theory set off a frenzied debate. Many scholars quibbled about this or that detail, while admitting that overall Rawls had successfully captured and capped two hundred years of liberal theorizing in one elegant formulation. A small but influential group of scholars insisted that Rawls had matters all wrong. They argued that Rawls crudely ignored the fact that our *identities*—race, class, sex, and religion—were precisely who we were. Sociologist Robert Bellah observed in the influential book *Habits of the Heart* that modern liberal culture of the sort Rawls and his acolytes advocated is more and more "a culture of separation."[51] Rawls's liberal philosophy is, by its nature, hostile to the very notion that a sense of community among like-minded people is a crucial part of living a good life. The most prominent of these critics were Michael Sandel, Amitai Etzioni, Michael Walzer, Charles Taylor, Alasdair MacIntyre, and Robert Putnam; collectively they were known as "communitarians." They argued (each in his own way) that unfettered rights-based liberalism undercuts the sort of group solidarity that makes rich ethnic and religious identity possible. Rights-based liberalism, they agree, is not a neutral position that equally facilitates all sorts of outlooks and identities. Quite to the contrary, liberalism of the sort that Rawls promoted was inescapably corrosive of the sort of traditional group identity demonstrated by Christian Scientists when they refuse medical treatment because their church advises them so—and, for that matter, Orthodox Jews in places like Kiryas Joel in New York, who wish to maintain segregated schools with public funds.

What Menachem Fisch offers is a third way to view these issues, one that navigates the shoals of Rawlsian liberalism and communitarianism. What those generally opposing views share, and Fisch rejects, is an axiomatic assumption that faith-based (or tradition-based or identity-based) communities pose a threat to one another that must be managed. Rights-based liberal approaches seek formal ways to ensure individual autonomy, thereby delimiting the influence anyone may exert over someone

---

50 John Rawls, *A Theory of Justice* (Cambridge, MA: Harvard University Press, 2009).

51 Robert N. Bellah et al., *Habits of the Heart: Individualism and Commitment in American Life* (Berkeley: University of California Press, 2007).

who views the world differently. Communitarian approaches seek to increase the autonomy of communities, thereby diminishing the influence of people from outside the community. Fisch suggests that the *other*, properly viewed, is not a threat to one's own view. The other enriches our own beliefs, and strengthens our own communities. However the two approaches differ, contemporary liberalism and communitarianism each seek to protect from the violent impact of the other. Fisch asserts that the other need not be viewed as a threat, but as an asset. In so doing, he sets the debate about tolerance on a new foundation.

Fisch had by now elaborated his theory of epistemic humility and confrontational theology, traced its long roots in the rabbinic tradition, and discovered its implications to extend far beyond the original epistemological questions that guided his scholarship, deep into the realms theology and politics. Along the way, a philosophical challenge arose; it was profound enough, and foundational enough, that, if unmet, some of Fisch's central philosophical claims might unravel.

The challenge was this. Expanding upon Popper, Fisch identified rationality with criticism. (Indeed, what made the eponymous rabbis of his book "rational" was their enduring search for opinions that contravene and criticize their own.) However, in order for criticism to meet its mark, it must be understood, absorbed, and internalized by the actor being criticized. For criticism to *improve* what we think and do, we must accept it, and transmute it into *self*-criticism. If we are to see the errors of our ways, we must make our own the criticisms set out by others. The criticism that others express towards our thoughts and actions become potent only when they are translated into our own criticisms of our thoughts and actions.

It is here that the philosophical challenge begins. For, as Fisch writes:

> To criticize is to pass normative judgment, for which a normative framework must necessarily already be in place... Conducting oneself rationally... is to hold oneself *accountable to* one's existing standards of propriety. But rationality requires of us not only to apply our standards *in* critical appraising but to subject them *to* critical appraisal, not only to act in accord with the standards and norms to which we are committed but to hold those very norms and standards in normative check. But how is that possible? How can the standards to which one is committed be effectively subjected to one's own rational appraisal if it is by means of those very standards that one rationally appraises?[52]

---

[52] Menachem Fisch, "Science, Religion, and Rationality: A Neo-Hegelian Approach," *Toronto Journal of Theology* 29, no. 2 (2013): 321.

In other words, all of us make sense of the world, and act in it, from within a given values framework. It is possible, of course, for someone who shares that framework to criticize something we think or do as illogical, inconsistent, or incoherent with our other beliefs or actions. But how can someone operating in a different normative framework ever apply criticism that hits the mark, especially about matters that are themselves normative? Wouldn't that be rather like a German speaker criticizing the syntax of our English speech, on the grounds, say, that we neglected to place our verbs at the end of our sentences? We would have no way to make sense of such criticism, much less to allow it to spur us to improve our communication.

This matters, of course, because if this challenge cannot be overcome—if there is no good philosophical account for how we can make sense of criticism offered by folks working within normative frameworks different from our own, and sometimes make it our own—then the powerful promise of epistemic humility that Fisch described in his earlier work would vanish. If the criticisms of Shammaites cannot be absorbed by the Hillelites, or those of the Christian pilgrims cannot be absorbed by Jewish Israelis, then the richly substantive pluralism that Fisch advocates—in which engagement with the other allows one to improve oneself—remains little more than a pretty dream.

Normativity and Rationality

It was to this challenge that Fisch devoted the essay in this volume "Science, Religion, and Rationality: A Neo-Hegelian Approach" and a fuller treatment in a book he wrote with philosopher Yizhak Benbaji, *The View from Within: Normativity and the Limits of Self-Criticism* (University of Notre Dame Press, 2011). The latter is a tour de force, engaging with vigor the great epistemologists and theorists of rationality of recent generations. (The power of the presentation, and its greatest pleasures, are in the details, most of which I cannot recount in this schematic account.) Fisch argues that is possible to bridge the gap between different normative frameworks, imperfectly but well enough for criticism to land. "What normative critics normally do," Fisch writes, "is frame their arguments ... from a perspective very close to that of their addressee, yet sufficiently (and subtly) different from it to be able to at least mount a valid argument."[53] One criticizing from another normative framework will impersonate the framework of the one being

[53] Ibid., 323.

criticized. Such impersonation is never perfect, and continues to convey criticism, yet it may be familiar enough to the one being criticized that its point may be absorbed. Linguists speak of "interim translations," in which not all the features of the original language have been massaged into the new language. Such "interim translations" may have the jarring effect of intensifying the attention of readers who understand the translation but register that it reflects a cognitive world slightly different from their own. Something like this happens, Fisch explains, when a trusted critical *other* tries to express her criticism in the normative framework of the one she seeks to criticize, as best she can impersonate it:

> Since the points of divergence between the two pictures—our own normative self-portrayal and that conveyed by our critic's argument—will pertain to the very norms her criticism calls into question, their incongruity, not unlike the case of a disturbing playback device, may well destabilize our initial commitment and render us *ambivalent* toward those norms. And norms to which we become ambivalent... lose their wholehearted volitional grounding, and with it their status as firm commitments. Once demoted, they will be subjected, as a matter of course, to the normative critical scrutiny of the self's surviving commitments. It is in this way, I submit, that exposure to the echo chamber of trusted normative criticism *is* capable of inducing the levels of inner discordance and self-alienation necessary for truly transformative normative *self*-criticism.[54]

Fisch's account is a philosophical solution to a philosophical problem, and that is how it is presented. It is worth noting that it is equally a human solution to the very human problem of how we can influence those whose worldviews differ from our own, and how we can be influenced by them. It is a problem that has always been important, but that in our day has become crucial. Ours is increasingly an age of *incommensurability*. It is an age in which we have grown skeptical of the very possibility of persuading those who view the world differently. In our day, the claims of climate scientists scarcely register on Republican politicians, and vice versa. The claims of advocates of GMO technologies are lost on their critics, and vice versa. The claims of monetarists mean nothing to Keynesians, and vice versa. We have lost our faith that people of one worldview can converse with people of another, much less influence them.

The notion of "incommensurability" is shared by those whose perspective is absolutist ("True Believers" of one sort or another) and those whose perspective is relativist. The absolutist is often persuaded that one who

[54] Ibid., 324.

has not accepted a revealed truth cannot understand it and is destined to be little more than a deluded detractor. The relativist often believes that one's truth is simply a preference, as mysterious and inexplicable as an affection for jazz or a taste for pistachio. The absolutist and the relativist share a fundamental belief that between people whose truths differ, a real meeting of the minds is impossible. It is this axiom that Fisch rejects.

As we have seen, Fisch has given a philosophical account of how such conversation can be accomplished and an account of why it must always be sought. These theoretical accounts he has also put into practice in rare and exquisite ways. He has traveled to Vatican City at the pleasure of the Holy See, where he told the assembled worthies that:

> By engaging in serious open and critical inter-faith dialogue ... with as significant and important other as the Catholic Church, ... the name of God is sanctified. It is sanctified not because we purport to bring to Rome His one true word, but because there is a very good chance that we shall be returning home from Rome better equipped for the ongoing task of working it out for ourselves.[55]

Fisch has also, for much of the past quarter century, run the annual International Theology Conference of the Shalom Hartman Institute, where he serves as Senior Research Fellow of the Kogod Center for the Renewal of Jewish Thought. The conference convenes each year Christian, Muslim, and Jewish theologians in Jerusalem, where they study together texts of each of the religions. As Fisch describes it:

> What we are ... fascinated by and concentrate upon and study and keenly discuss with each other are each other's *intrafaith* dialogues.... The idea is not to search for shared ground or for common denominators, and the idea is certainly not to convince each other of the truth of one's own faith by proclaiming it. On the contrary, the idea is to think aloud and deliberate major issues in one's own religious world in the company of the religious *other*. To expose oneself religiously, in the presence of the religious other, precisely because of and in order to benefit from, his or her *otherness*. There is no long or short term aim of overcoming these differences, and of seeking to expose the one shared religious truth. We gather to enjoy and to celebrate our differences by creating a fascinating polyphony, of seriously engaged religious voices, witnessing and questioning and learning from each other. Bringing to each other exactly what no one and no faith tradition can give

55 Fisch, "Judaism and the Religious Value of Diversity."

> itself, namely the gift of a truly different perspective.... We aim... to create a religious solidarity of differences.[56]

Something of the same spirit animates the Center for Religious and Interreligious Studies (CRIS) at Tel Aviv University, which Fisch founded in 2010 and has directed since. The Center, in collaboration with the Cambridge University Project for Religions in the Humanities and a group of senior scholars at the Goethe University, Frankfurt, aims to establish a new field, Interreligious Studies, that will find its place alongside the established fields of Religious Studies and Comparative Religions. The field, Fisch has said:

> is being built on the understanding that to properly understand the three Abrahamic faiths, one has to take into account the dialogue that each is conducting with the other two traditions. We didn't evolve separately, and we don't exist independently. We are all part of the same, interconnected Abrahamic tapestry. It's therefore essential that they be studied and taught within the same institutional framework, not as three separate fields of study. This is what makes our initiative unique.[57]

CRIS in establishing a graduate program in Interreligious Studies, recruiting faculty, executing international conferences, sponsoring public lectures, and preparing to launch a journal, is, in a sense, a materialization of Fisch's scholarly work, a translation of his ideas into institutions.

Concluding Overview

Thirty years have passed since Menachem Fisch completed his doctoral dissertation, and the course that his scholarship has followed in that time could not then have been anticipated. How did a philosopher and historian of science, with an interest in epistemology and a fascination for Victorian science, come to be one of the most creative and important Jewish thinkers of the late twentieth century and early twenty-first century? On the surface, it is an unlikely story.

---

[56] Menachem Fisch, "Shalom Hartman Institute's Theology Conference Explained," https://www.youtube.com/watch?v=Eix5sut4OEg (accessed November 1, 2015).

[57] "A Pan-Abrahamic Dialogue," https://english.tau.ac.il/interreligious_studies (accessed November 1, 2015).

As it happens, the tools of the philosopher and historian of science, and the *concerns* of the philosopher and historian of science, proved especially useful for addressing intellectual, religious, political, and social challenges that Jews face in our times. They drew Fisch to consider the discontents both of foundationalism and of relativism. They fixed his attention of the vexed question of what distinguishes rational from nonrational action. They drew him to the ask how people embedded in a worldview—scientists of a given theoretical orientation, or religionists of a particular persuasion—can understand and absorb views from outside that worldview, as they must if what they think and act are ever to develop, progress, and improve. They drew him to ask how the seeming *incommensurability* between the different beliefs held by different people in different communities of belief can be overcome. They drew him to ask how people who believe one thing can ever converse, profoundly and productively, with people who believe something else. These were questions that a philosopher and historian of science could hardly escape in the most tumultuous and productive period the field has ever known, the period in which the predominance of logical positivism waned and new sorts of explanations for science were floated by a very varied collection of theorists such as Karl Popper, Thomas Kuhn, Imre Lakatos, Paul Feyerabend, Michelle Foucault, David Bloor, Barry Barnes, Margaret Rossiter, Donna Haraway, Bruno Latour, and a great many more.

These same questions proved to be among the most important and vexing questions of our times, and not just for science. We have witnessed a disintegration of the very notion of public discourse. Political scientists warn, in ominous tones, of crises of legitimation, crises of governance, crises of public trust, crises of public confidence, crises of leadership, crises of authority, crises of fragmentation, crises of representation, crises of multiculturalism, crises of liberalism, and more. "Only connect!," E. M. Forster exhorted in 1910, and by our day our powers to do so are dangerously diminished. So, too, our talents for living peaceably and constructively with those who are, to us, the *other*.

All this has had its own, particular impact upon Jews. Although the past century has been one of great creativity in Jewish spiritual and intellectual life, it has also seen renewed splintering of Jewish life. Denominations multiplied: Ultraorthodox, Orthodox, Modern Orthodox, Conservative, Reform, Reconstructionist, and Renewal are rubrics under which one can frame a complex taxonomy of subgroupings. Substantive exchange between these groups is rare, and rarely constructive. Recent decades has seen a rise of Jewish *neoparochialism* and an uptick in what might be called *exclusionary* Judaism. Separatist ultraorthodox communities thrive in Jerusalem and Bnei Brak in Israel, in Williamsburg and Kiryas Joel in New York, and in Antwerp and Strasbourg in Europe. In Israel, President Reuven Rivlin recently warned that "a new order" has emerged, characterized by tribalism:

> Children born in the State of Israel are sent to one of four separate education systems.... A child from Beth El, a child from Rahat, a child from Herzliya and a child from Beitar Ilit—not only do they not meet each other, but they are educated toward a totally different outlook regarding the basic values and desired character of the State of Israel. Will this be a secular, liberal state, Jewish and democratic? Will it be a state based on Jewish religious law? Or a religious democratic state? Will it be a state of all its citizens, of all its national ethnic groups? Tribe, by tribe, by tribe, by tribe. Similarly, each tribe has its own media platforms, newspapers they read, television channels they watch... In the State of Israel the basic systems that form people's consciousness are tribal and separate, and will most likely remain so.[58]

The difficulties that Israelis have in speaking with one another are part of a problem that is larger still and has to do with the dramatic renewal of Jewish Sovereignty. The unending and vexed debate over how to balance the country's Jewish and democratic pretensions is one sign among many that Israel has not found a way fully to accommodate religious "others." This problem is ramified by Israel's military control of the West Bank and the more than two million Palestinians who live there.

Something similar is true on a still-larger scale. The ability of Israel's leaders to communicate with the rest of the world has lately deteriorated. The country's actions seemingly make less and less sense to many beyond its borders. At the same time, the actions of other states grow less intelligible to many Israelis. With alarming frequency, the actions of others are attributed to anti-Semitism. But the claim that this or that diplomatic development was motivated by anti-Semitism is often, in effect, a claim that it has no intelligible explanation whatsoever; that it is motivated by a hatred of mysterious, perhaps mystical, origins, and that it is therefore not given to rational analysis. It is for this reason that, as Israel grows more isolated, fewer Israelis believe that it is within their power to relieve their growing isolation. Zionism in our day, and Judaism more generally, are

58 "President Reuven Rivlin Address to the 15th Annual Herzliya Conference," http://www.president.gov.il/English/ThePresident/Speeches/Pages/news_070615_01.aspx (accessed November 1, 2015).

suffering from a crisis of communication so fundamental that the future thriving of Israel, and of Judaism itself, are in doubt.

The power of Menachem Fisch's scholarship owes, in part, to the light it sheds on this crisis, and how best to address it. Over decades of meticulous philosophical, historical, and theological research, Fisch has described rich Jewish religious and intellectual resources with which the crisis can be met. Further, he has shown how these resources can be made to facilitate profound engagement among Jews of diverging beliefs, between Jews and believers of other religions, between Jewish and non-Jewish Israelis. and between Israelis and the rest of the world.

In all this, Menachem Fisch has, in a way, continued work begun by his grandfather, Rabbi Dr. Solomon Fisch, and his father, Professor Harold Fisch. The journey of Solomon Fisch—from the relatively closed world of a Jewish town in Eastern Europe so unitary in its attitudes that a single *apikores* remained a vivid memory decades after his death, to the cosmopolitan world of German Orthodoxy, to the quotidian Jewish nonobservance of the English midlands—left him with little doubt that the emancipation of Jews in the modern West, and their assimilation into Christian society, could have corrosive effects on their adherence to Jewish tradition. Solomon Fisch addressed this challenge with sophistication but not, by his own measure, with success. Much like his mentors in Germany, advocates of *Torah im Derekh Eretz*, he was alive to the cultural and intellectual richness of the (non-Jewish) West. Books in varied languages mingled on his shelves, scholarship of varied provenance was absorbed into his own research; he was at once a *talmid hakham* and a learned scholar. Still, in the end, for Solomon Fisch, the greatest challenge faced by Jews of his day was to gain fluency in Western culture *without* being unduly influenced by it. It was to *prevent* interaction so penetrating as to be corrosive of *Yiddishkeit*. This was a challenge of existential importance. It was for this reason that, while still a young student in Germany, Solomon Fisch recoiled from those he met there who embraced German culture in a deep way. Indeed, for Solomon Fisch, *Torah im Derekh Eretz* was an ambivalent program in which engaging the outside world was also a way of holding the outside world at bay. He believed that it is possible and worthwhile to draw from the world *without*, in so doing, challenging what he took to be the essence of Judaism. *Kultur* and *Bildung* could live side by side with a pious orthodoxy. What they offered could enrich the Orthodox Jew, without challenging his Orthodoxy.

It may well be that, during his decades in England, Solomon Fisch came to doubt the sustainability of this approach. There he witnessed a diminution of *Yiddishkeit*, although in this Midlands this came as a result of crasser passions (for money and entertainments, for instance) than the passion for high culture he saw among Jews in Germany. One senses in Solomon Fisch, in his later writing, a disappointed awareness that the gap between the lax *Yiddishkeit* he witnessed in England in his last decades there and the taut *Yiddishkeit* of his youth owed the overly enthusiastic openness of Jews in his day to the ways of their Christian neighbors.

Harold Fisch, too, was alive to the problem that vexed his father and, in the end, he found answers similar in spirit to those his father found. His facility with some of the greatest writing the West ever produced in English was without parallel (a typical review of his book on Shakespeare, Milton, and Blake found in it "wisdom and reflection, from deep knowledge and long thought, ... all driven by the inexorable logic and importance of his insights into these three great writers"),[59] yet he was confident that, while this literature might *illuminate* his Judaism, it was not within this literature to *alter* his Judaism in any way. (His answer to the question of whether his study of Western literature could "compromise" his "Jewish beliefs and practices" was a hearty "Quite the contrary!") This same confidence was a foundation of his Zionism ("Whether we were in Poland or Germany or England," he wrote, roughly tracing his father's path of emigration, "we were a people in exile." To believe a Jew could thrive in such places "was to court ruin and disaster.")[60] Harold Fisch was alert to Western culture and alive to its beauty and richness. But, like his father, he sought ways to ensure that it did not compromise his Jewish beliefs and practices. Among these was his avid Zionism by which, he wrote, we are "called upon to stand apart, to confront the world in the name of values peculiar to the Jewish tradition."

It comes as no surprise that Solomon Fisch and Harold Fisch faced similar challenges. Nor is it surprising that, in his day, Menachem Fisch came to address them, as well. The question of how Jews, and Judaism, ought to find their place within and among the nations is the great question that has absorbed Jews (and many others) since the first portents of emancipation.

---

[59] Rose Arnold, "Review of *Radical Religion from Shakespeare to Milton: Figures of Nonconformity in Early Modern England* by Kristen Poole; *Shakespeare and the Bible* by Steven Marx; *The Biblical Presence in Shakespeare, Milton, and Blake: A Comparative Study* by Harold Fisch," *Literature and Theology* 16, no. 1 (2002): 104.

[60] Harold Fisch, *Who Knows One?* 54–55.

Menachem Fisch has provided an answer that is far more radical than that offered by his father and grandfather. They proposed to preserve Jewish thought and practice by fortifying walls around it (and within it, dividing different denominations). He proposes to enliven Judaism by doing the opposite, by laying it open to the influence of others, by seeking to engage, profoundly, in critical discourse with others. The results of this approach are far-reaching, as we have seen. They include new notions of Jewish sovereignty, new models of Jewish engagement with other Jews, with people of other religions, and people with no religion, new approaches to Jewish learning, and much more. Yet, for all its radicality, the Judaism Fisch has proposed is one based in a tradition that is alive in canonical texts, if long undervalued or overlooked.

At the heart of Menachem Fisch's work has been an appreciation of the crucial importance of epistemic humility, and of the indispensable role others play in our own efforts to increase the wisdom and decency of what we think and do. His writing and teaching have reflected the modest tentativity that he advocates for others, as well as a constant yearning to draw others into discussion and to hear their criticisms. It is perhaps owing to this that he has produced scholarship of such sweeping reach and, in our troubled times, enduring importance.

RATIONAL RABBIS: ITS PROJECT AND ARGUMENT

Menachem Fisch

Rational Rabbis[1] aspires to make two main points, one philosophical and contemporary, the other interpretative and historical. The book's philosophical undertaking, presented in Part I, is to develop a central insight of Karl Popper's into a fuller theory of rational endeavor. The book's interpretative and main undertaking, presented in Part II, is to argue (a) that the talmudic literature bears clear witness to a tannaitic view of humanly possible intellectual achievement intriguingly akin to the theory of rationality proposed in Part I, and (b) that despite appearances to the contrary, this voice is centrally responsible for the Bavli's halakhic discourse and project. The TR session at AAR 2002 focused on this last claim by means of a close reading of the '*meitivi' sugya* presented in Bavli, *Berakhot* 19b–20a. What follows briefly summarizes that reading and outlines its broader philosophical and hermeneutical settings.

1. Although he is one of the better known and most widely read thinkers of our time, Popper's impact on professional philosophy has been marginal. He is widely regarded as a more-or-less effective, if semi-positivist combatant of positivism who retained little relevance in the heady aftermath of the works of Kuhn and the later Wittgenstein. *Rational Rabbis* insists that Popper's "constructive skepticism" represents much more than a mere first step toward these more robust and better developed critiques of the conceits of modernism. Popper was primarily a philosopher of science, but I am less interested in his theory of science *per se* than in the general view of rational endeavor that this theory implies. Like many of his critics, I find his realism unfoundedly naive, his attitude toward language wholly inadequate, his portrayal of science generally simplistic and his notions of corroboration and verisimilitude largely wrongheaded. But I do find his basic fallibilism crucially important and attempt to develop it further. It is the idea that the rationality and advancement of science owe far less to a

---

* This article was first published as: Menachem Fisch, "Rational Rabbis: Its Project and Argument," in *Journal of Textual Reasoning* 4, no. 2 (2006). Reprinted with permission from the *Journal of Textual Reasoning*.

[1] Menachem Fisch, *Rational Rabbis: Science and Talmudic Culture* (Bloomington: Indiana University Press, 1997).

confident reliance on its data, methods and warrants, than to the self-doubting Socratic 'dialectic of interrogation' to which they are regularly subjected. Incapable (as a matter of logic) of objectively confirming their efforts, let alone of proving them, scientists, argues the Popperian, can, in principle, boast no more than prudently to have subjected them to the most thorough tests at the scientists' disposal. From this follows the identification of the rational with the critical.

Popper's fallibilism is a form of skepticism, but not of the wholesale, paralyzing Humean variety. Although serious criticism begins with doubt, criticism and doubt have little in common. Entertaining a doubt adds up to little more than applying a question mark, or raising one's eyebrows; serious criticism, by contrast, requires fashioning an argument. To doubt is to suspect something might be amiss, to criticize is to *argue* that it is. Skeptical discourse requires a supply of interrogatives, critical discourse requires rich background knowledge and a developed logic of problem-seeking and solving. Criticism necessarily presupposes doubt, but is also a necessary prerequisite for positive action. In the face of suspected imperfection the first step toward improvement will always be critical. Hence the term 'constructive skepticism'. The theory of rationality outlined in the first part of *Rational Rabbis* does not fall ready-made from Popper's pen. Rather, I find his exposition lacking in two main respects. First, although Popper wrote widely, he had surprisingly little to say philosophically about the central notions of his system: criticism, rationality, problems and the processes of problem seeking and solving. They are discussed to an extent within the framework of his theory of science, but not in general. Nowhere, for instance, is the difference between criticizing and merely doubting analyzed. Nowhere does he stop to ask what must be assumed uncritically for criticism to be effective. And for all the stress he laid on problem-seeking and problem-solving, he offers no general, philosophical account of problems or solutions—all of which are essential for a full-fledged philosophical elaboration of his views.

The other major shortcoming of Popper's work is the way it conceives the opposition. Motivated by the dramatic developments in physics during the first quarter of the twentieth century, Popper fashioned his anti-foundationalist vision of science to combat foundationalist tendencies in the philosophy of science, especially those of the positivists. This foundationalism boasted an articulation of real empirical scientific accomplishment devoid of an uncriticized empirically given foundation. Today, foundationalism as such is, from a serious philosophical perspective, a matter of the past. The groundbreaking works of Sellars, Wittgenstein and Quine during the 1950's and 60's have rendered it a philosophical myth—'The Myth of the Given', as they dubbed it. The most serious challenge to any general theory of scientific rationality and progress issues now from radical relativist, rather than foundationalist quarters—a relativism that hinges on the uncriticizable given-ness of language. Latter-day relativists argue forcefully that the language games (conceptual schemes, paradigms, or 'final vocabularies') within, or with which we operate, generate forms of life, cognition and perception that are, like Kant's categories and forms of intuition, in principle immune to criticism from within. The fact that, unlike Kant's categories, language games, vocabularies and paradigms, along with the forms of life they generate, are both humanly crafted and boast histories makes no difference. Critical reflection is flatly denied any role in their development. Kuhn put it bluntly—describing for the sciences what Davidson, Rorty and Brandom would argue rigorously for all manner of human conceptualization—paradigms are either unreflectively presupposed and applied, or abruptly replaced by alternatives sufficiently "incommensurable" as to render the new incomprehensible in terms of the old. Only in retrospect can the new paradigm be comprehended as a solution to problems harbored by the old. Writ large, this adds up to saying that a language (in the broad, Wittgensteinian sense of the term) cannot be troubleshot by those who speak it. To the best of my knowledge neither Popper nor any of his followers have so far made any attempt of philosophical significance to combat this so radical of latter-day challenges to rationality.

Part I of *Rational Rabbis* proposes a general characterization of critical discourse designed to lend philosophical generality to Popper's basic insight in a manner capable of meeting this challenge. It does so by means of the general notion of a *goal-directed system*. A goal-directed system is any structure, theoretical, institutional or material, designed or adopted by human agents as a means to specific ends. The notion of such a system serves as a teleological template for defining and analyzing all manner of contrived troubleshooting. In the broadest possible sense of the term, any "problem" can be defined as the failure of a particular system to achieve fully one of its intended goals; a discrepancy, that is, between what it is in fact capable of accomplishing, and what it is meant, desired, or expected to accomplish. Problems are thus construed ontologically. They are features of systems about which human observers may be right or wrong. Systems that require our attention (as opposed to those we apply unthinkingly) are those we regard as susceptible to critical appraisal and capable of improvement. They are the systems we suspect harbor problems.

A modified goal-directed system will be said to have been *improved*, enhanced, or to have undergone *progress* exactly if problems it had harbored are now wholly or partly solved. (Conversely, a system will be said to have *deteriorated* or *regressed* exactly if it is now more problematic than it was.) Solutions, and by implication the measure of progress they represent, are thus also objective features of goal-directed systems about which one may be right or wrong. Such a general account of problems and problem-solving allows one to retain a general Popperian rendition of progress without having to commit oneself to Popper's particular view of the aim of science and the nature of scientific problems. It also opens the way for an equally general definition of rational endeavor.

To act rationally, I propose, is to act upon an allegedly problematic goal-directed system with a view to improving its performance; to apply one's brains consciously and deliberately to the task of exposing the problems the system harbors, with a view to solving them. To act rationally, in other words, is *to contrive deliberately to effect progress* in the sense described. Needless to say, success in this regard is not a criterion for deeming an action rational. (In requiring that rational action be conscious and deliberate—that is to say, by insisting on agency—and in allowing rational actors to err and fail, it is easy to see that such a view of rationality is quite at odds with currently popular game-theory models of rational choice and action.)

Part I of *Rational Rabbis* pays its debt to the work of Popper and his students by setting forth and arguing against the backdrop of the recent history of philosophy of science. But the theory it proposes is not a theory of scientific rationality or of scientific problem-solving per se. It purports to make a point about the nature of human accomplishment and of human rationality in general. It presupposes little more than an elementary epistemic humility few could deny should apply to a vast majority of human enterprises. A theory of rational action should be as philosophically lean as possible so as to allow for philosophy itself to be pursued rationally. Not all of my critics have appreciated the generality of my philosophical argument, and have taken me to be arguing for some sort of direct and special affinity between the nature of religious and scientific thought. That was not my intention. What I have tried to do is to argue from science to a general theory of rationality and from that back down to all manner of enterprising activity—those of a legal or even religious nature included. Thinking in terms of goal-directed systems in general allows one to view, say, a baseball player working methodically to improve his swing, a poet to improve a verse, a mechanic to fix a car, or a legislator to amend the law, as all doing the same sort of thing as a researcher mixing batch after batch of experimental vaccine. It allows one to view them all as acting rationally to lay bare and solve their own very particular, and very different, problems. Now, it goes without saying that specific *halakhot*, whole bodies of halakha, methods of halakhic reasoning and interpretation, meta-halakhic principles, etc., can all be described as goal-directed systems (some nested, some independent).

As opposed to the normative aspiration of Part I, the thesis presented in Part II of *Rational Rabbis* is purely descriptive. My personal preferences notwithstanding, my thesis in Part II is *not* that halakha can, or should, be treated as potentially revisable and, therefore, as subject to the type of rational troubleshooting outlined above. The thesis I argue for is that the talmudic texts bear clear witness to the existence of a rabbinical school of thought that clearly seems to have viewed the development of halakha along such lines.

2. More precisely, Part II aspires to establish and then to explain two interpretative claims. The first is the relatively straightforward claim, argued for in some detail, that a variety of sources, of authentic and alleged tannaitic origin, bear clear witness to a major dispute regarding the required attitude of the halakhic authorities toward the halakhic tradition. For lack of a better terminology I dub the two parties to the debate 'traditionalist' and 'antitraditionalist'. Traditionalists insist that the halakhic tradition—i.e. the body of existing halakhic rulings at any one time—is at all times binding, absolutely binding. Their position is formal. Former halakhic rulings are binding not because they are found just or reasonable, but because once a ruling is made it is irrevocable. According to Tosefta, *Sanhedrin* 7:1, in which the traditionalist position is stated most clearly, no halakhic institution, not even the Great Sanhedrin itself, has the authority to overturn an existing halakhic tradition. Halakha develops and expands, but does so exclusively by accumulation, never by revision. From a traditionalist perspective, halakhic innovation is hence wholly limited to genuine lacunae. And once a vote is taken, the new matter is considered settled once and for all. Antitraditionalists, by contrast, consider it the religious duty of the halakhic authorities not only to transmit and to occasionally add to the system, but to hold it in constant review and to revise it as they see fit.

A comparison of the Mishna and Tosefta versions of the first chapter of *Eduyot* provides a good example of how the dispute plays itself out in the tannaitic literature. Both renditions ask why minority opinions should remain on record despite their rejection. Both record two answers: one markedly traditionalist and one markedly antitraditionalist. But the two texts decide the matter differently, with the Tosefta siding forcefully with the traditionalist option (claiming that the reason for keeping records of rejected opinion is to make sure they remain rejected! They are remembered, that is, in order to remain forgotten!) and the Mishna taking an equally forceful antitraditionalist stand (ruling that minority opinions are to remain on record, as a resource for future *Batei Din* who may see fit to overturn their predecessors' majority rulings and revive the ones rejected!)

It is not hard to see why I consider halakhic antitraditionalism a position closely akin to the neo-Popperian view of rational endeavor proposed in Part I. Unlike other forms of religious discourse (which are left by the talmudic literature decidedly undecided!), halakhic discourse demands closure—at least with respect to practical matters. Antitraditionalists treat all halakhic decision-making as tentative and revisable (by the appropriate authorities, of course). Even the great Sages of old are liable to have erred or to have failed to see the larger picture, let alone to have not anticipated future changes of circumstance and opinion. And if they are considered fallible, *al ahat kama ve-khama*—all the more so—that we are! Talmudic antitraditionalism is hence motivated by a deeply religious sense of responsibility for the constant development of the Oral Torah, coupled to an equally deep religious epistemic humility. Whether or not one can talk of a meaningful notion of halakhic verity remains an open question in antitraditionalist discourse.[2] All anti-traditionalists agree that even if one could speak of halakhic verity, halakha would remain *humanly* unverifiable. It may be tested and questioned, however. Hence the Mishna's explanation of why halakhic rulings of the past should be transmitted along with the debates that preceded them. Read thus, Mishna, *Eduyot* can no longer be taken to mean, as it is commonly understood, that future (re)rulings are, therefore, *restricted* to minority positions that were formerly voiced. Indeed, in the three preceding *mishnayot*, the Mishna itself (quite deliberately, I believe) cites the differing opinions of Hillel and Shammai with regard to three separate issues, and rules explicitly on each of them *lo ke-divrei ze ve-lo ke-divrei ze* (neither according to the one nor according to the other!)

Still, with the possible exception of tractate *Eduyot* itself and some of the Bet Shammai and Bet Hillel disputes, the vast majority of the tannaitic material bearing on the traditionalist-antitraditionalist debate is declarative and theoretical. In this respect the two *talmudim*, especially the Bavli, have much more to offer. Unlike any antecedent tannaitic document, they are fashioned as ordered series of detailed, blow-by-blow responses to an earlier, authoritative halakhic text, the Mishna. They ponder and attempt

---

[2] For a fascinating rejection of the idea of halakhic truth see Rashi: Bavli, *Ketubot*, 57b s.v. *mai ka-mashma lan*.

to reconstruct the complex lines of reasoning that led their predecessors to their conclusions, and continue vigorously from where they left off. The narrative forms adopted in their discursive portions resemble those of an elaborate protocol, which comes accompanied by a clear and dispassionate running commentary. The Bavli especially leaves a strong impression of being primarily a didactic work, designed to convey a carefully crafted vision of *talmud-Torah*. If one seeks evidence of self-aware talmudic notions of halakhic development *in operatio*, it is here, in the Bavli's narratives of deliberation and dispute, that one should look.

3. The kind of amoraic *sugya* around which one would expect traditionalist and antitraditionalist redactors to differ most conspicuously are those that focus on apparent inconsistencies straddling the Talmud's great generational divide, between tannaitic and amoraic halakhic ruling. These are cases, commonplace throughout the Bavli and Yerushalmi, in which a well-formed amoraic position is confronted with tannaitic sources that imply the opposite. For traditionalists, of course, the latter-day amoraic view will always be the one deemed problematic. For antitraditionalists, the direction of questioning may, in principle, go either way, but will normally go from new to old, from amoraic to tannaitic, rather than the other way round. Antitraditionalists are expected to form their halakhic views by seriously troubleshooting those they inherit. Once they arrive at a decision, and pronounce their ruling, they will naturally deem it superior to those they had criticized. Thus, when earlier and later positions are confronted, it will usually be the former who needs to answer the latter. Of course, the framer of the *sugya* (henceforth the *stam*) might find the later amoraic position inherently puzzling, but in such cases one would expect him to question the amora's reasoning rather than confront him with an earlier source. Representatives of the two schools, therefore, will be expected to conduct and narrate their transgenerational confrontations very differently. However, even here, the talmudic text is far more equivocal than one would expect.

The Bavli introduces such confrontations by means of the word "*meitivi*" ("a response was offered", "an objection was raised"). The very word *meitivi* would seem to favor a traditionalist approach, for it implies that an amoraic position is called into question merely by the fact that a tannaitic source appears to suggest the contrary. The Bavli contains six-hundred-fifty or so *meitivi* confrontations (many of them comprising more than one challenge). In a little less than 8% of them the amoraic position is declared "problematic" (*teyuvta*) and in 3.8% of the cases, seriously so (a 'double' *teyuvta*). The opposite, though, seems never to occur. Nowhere is a tannaitic statement ever outwardly declared refuted or rejected as the result of a *meitivi*-type confrontation. When discordant amoraic and tannaitic statements are confronted, the Bavli appears to adopt a clear traditionalist stance; always deeming the former potentially at fault and in real danger of being rejected.

But does the inner logic of such negotiations always lend itself to the marked traditionalist idiom of the way they are narrated? A reasonably reliable method for determining a person's understanding of a problem is to look at the way he or she elects to solve it. It stands to reason that the system deemed questionable will usually be the system that is subsequently modified, limited or cast aside. (The remedy, in other words, is usually applied to the wound.) And it is here that one can frequently detect a disparity between what the Bavli says its doing and what it actually does—for almost as a matter of course it is the tannaitic, rather than the amoraic text that ends up suffering the consequences of the *meitivi* challenge. Contrary to the traditionalist rhetoric of *meitivi*-type objections, the ways in which they are met and dealt with in practice often suggests a more antitraditionalist mindset.

However, *Rational Rabbis* does not stop here, but argues further that the discrepancies between the form and content of many of the Bavli's narratives of transgenerational discourse are contrived and deliberate. The Bavli, I claim, in its halakhic discourse, contains a clear and central, yet deliberately disguised, antitraditionalist voice doing antitraditionalist halakhic work. And it is here that the *meitivi sugya* of Bavli, *Berakhot* 19b–20a, studied at the AAR 2002 TR session, enters the story. I devote more than forty pages of *Rational Rabbis* to this one remarkable unit of talmudic discourse as I believe it represents the Bavli's own paradigm of a *meitivi*-type confrontation; This *sugya* is the Bavli's textbook example, as it were, of transgenerational halakhic give and take. The *sugya* is conducted, I believe, for the benefit of its more advanced readership, with a view to deliberately displaying the determined antitraditionalist project operating behind its carefully maintained traditionalist rhetoric. And yet this is achieved without the *sugya* itself ever relaxing its own deliberately misleading traditionalist rhetoric! Unless I am grossly mistaken, this is one of the most ingeniously crafted units of discourse in rabbinic literature.

4. At first blush the *sugya* appears almost boringly commonplace. A general halakhic ruling, attributed by R. Yehuda to his teacher, Rav, is cited and challenged by five different tannaitic sources that all clearly suggest the contrary. Each source is in turn either reinterpreted or severely limited so

as to harmonize it with Rav's later position. It is the type of *sugya* seasoned learners skim through almost thoughtlessly. There seems nothing unclear or unusual about it—nothing, that is, until one takes a second look.

A second look discloses a series of peculiarities whose joint effect I find quite dramatic. These reveal themselves at four levels of reading: first, when the *sugya* is considered against the backdrop of the Mishna on which it is commenting; second, when the details of its actual narration are examined; third, when the wider implications of Rav's ruling are noticed; and fourth, when the *sugya* is read in conjunction with the two units of discussion that immediately precede and succeed it. But before summarizing the evidence, let me briefly state to what I think it attests. It attests, I believe, to a remarkably subtle, yet powerful attempt on behalf of the *stam* to explode the traditionalist rhetoric he employs and to proclaim a pronounced antitraditionalist view of halakha. He is not merely declaring Rav's particular view superior to the tannaitic tradition in question, but is urging upon his more seasoned readership the understanding that, rhetoric notwithstanding, a keen, self-conscious antitraditionalism is, in truth and in general, the name of the (his) game. And all this is achieved inaudibly and indirectly, while hardly missing a beat in the normal, surface narration of the *sugya*. (Needless to say, that in presenting his antitraditionalism almost inaudibly, the *stam* is teaching his advanced readers, secondarily or even unwittingly perhaps, also always to be careful to conceal their true colors!)

5. Let us now look briefly at each of these four levels of reading. Rav's ruling comprises a specific prohibition presented as illustrative of a general halakhic principle:

> R. Yehuda said in the name of Rav: If one discovers mixed kinds in his garment, he should take it off even in the street. What is the reason? [it says:] "There is no wisdom nor understanding nor counsel against the Lord" (Prov. 21:30); wherever a profanation of God's name is involved no respect is paid (even) to a teacher.

According to Rav, then, considerations of respect should under no circumstances take precedence over one's immediate religious duties—not even respect for one's teachers. This is followed immediately by five *meitivi*-type challenges leveled from *beraitot* that imply that certain religious duties are, and should be, deferred in order not to offend others. This discussion of Rav's dictum is located two pages into the Bavli's commentary on Mishna, *Berakhot* 3:1. What is striking about this is that Rav's principle seems clearly to contradict the very *mishna* on which it comments! The *mishna* states explicitly that "One whose dead (relative) lies before him," as well as those who later actively participate in the funeral, are all "exempt from the recital of the *shema*, and from the *tefillah*, and from *tefillin*, and from all precepts laid down in the Torah." And the same goes for the *beraita*, cited on 19b as the cue for introducing Rav's statement, that rules in line with the Mishna that, among other things, those who attend a funeral "on account *of the mourner* are exempt, but those who come for their own purposes are not exempt." Neither source offers an explanation for the exemption. Nor does the Bavli ask for one, but it does do all in its power to imply that the reason is respect for the dead and for the mourners. For those who stop to ask themselves why the *mishna* and *beraita* rule as they do, and to ponder the relevance of Rav's dictum to their discussion, the discrepancy between them is glaring.

What makes the discrepancy all the more glaring, however, is the fact that it is passed over in complete silence. Normally, when something like this occurs, commentators seek for an explanation they feel the *stam* would have considered so obvious as to not merit mention. This won't work here, however, because Rav's alleged departure from tannaitic halakha is dealt with in detail, with the *stam* proceeding immediately to confront it with a series of five tannaitic sources that appear to contradict him quite clearly, and of which three explicitly concern respect for the dead and for mourners. As one would expect, all five are met in customary fashion by radically reinterpreting each of them so as to leave Rav's dictum intact. All along the unresolved and glaring discrepancy with the Mishna and *beraita* is allowed to lurk untreated in the background. I say 'allowed' because allowing for this discrepancy has to be deliberate. Such a glaring incongruity cannot be due to mere oversight.

At this deeper, more subtle level the *stam* seems to be calling his own bluff. As hard as he may try, he must know that his less naive readers will not be able to ignore or forget the Mishna or *beraita*. And the greater his effort to harmonize Rav's statement with other apparently conflicting sources, the harder felt will be the total absence of a similar treatment of the Mishna and *beraita*. I find it extremely hard to believe that the framer of the *sugya* might have been unaware of his omission, and suggest that both the build-up of the tension and the fact that it is passed over in utter silence are quite intentional. They serve, I believe deliberately, to subtly lay bare the intentional superficiality of the entire harmonization project. Indeed, the very fact that Rav's statement survives the *meitivi* confrontations with the five other tannaitic sources almost intact would suggest that, in the *stam*'s opinion, the Mishna is not merely contradicted, but is actually superseded by Rav's ruling!

6. The second unusual feature of the *meitivi* confrontations of *Berakhot* 19b is the unique manner in which they are introduced. Nowhere else in the Bavli does a *meitivi*-type objection come accompanied, as it does here, by a running commentary in which the *stam* explains which of the two texts he finds puzzling and why. *Meitivi*-type confrontations elsewhere are always introduced by the exact same format:

> Amora *A* says *X*; *meitivi*: citation of a tannaitic source implying the negation of *X*.

Typically the *stam*'s only contribution of his own at this stage (other than juxtaposing the two sources) is the word "*meitivi*" itself. The format implies that, given the obvious incongruity between the amoraic and tannaitic statements, the problem is obvious. How one understands it will depend, however, on one's approach. As noted at the outset, for traditionalists any conflict between amoraic and tannaitic rulings will automatically constitute a problem for the former. His question will always be: "how can the amora be saying what he appears to be saying, if the tannaitic source in hand states what it appears to state?" By contrast, the challenge an antitraditionalist would encounter in the course of such a confrontation would have to do with evaluating motives and reasons. Since for him latter-day positions are expected to be outcomes of serious criticism of earlier ones, it is only natural that the earlier, tannaitic source frequently will turn out to be the more puzzling of the two. An antitraditionalist will state his question thus: amora *A* claims that the halakha should be *X* for reason *R*. Curiously, the tannaitic text in hand implies the negation of *X*. Why, as seems to be the case, did the tannaitic authors not think of *R* or deem *R* inappropriate? Only an avid antitraditionalist will be able to seriously formulate a *meitivi*-type query in this fashion. And this is precisely the way the framer of *Berakhot* 19b scrupulously describes the puzzlement entailed by each of the five *meitivi*-type confrontations he conducts! Of each of the five beraitot he asks the same: "Why (is this so)?" Why doesn't the *beraita* say (as does Rav) that "There is no wisdom nor understanding nor counsel against the Lord" (and hence: wherever a profanation of God's name is involved no respect is paid (even) to a teacher)"? Unless we interpret the *stam*'s formula as a rhetorical, somewhat sarcastic challenge to Rav—a reading I find highly inappropriate—it is hard *not* to read it as a clear, if fleeting disclosure of antitraditionalist bias.

In addition to the antitraditionalist gist of the *stam*'s comments, the insistent, case-by-case, chant-like repetition of the identical interrogative formulae serves to create the distinct impression of a schoolroom example of which I spoke above. If one studies *Berakhot* 19b attentively, with some knowledge of similar *sugyot*, one has the distinct feeling of being patiently, very patiently taught to what a *meitivi* objection amounts, and how to pose, and deal with it. The lesson, however, will be completely lost on two kinds of reader: the less knowledgeable students who simply fail to register the uniqueness and significance of the *stam*'s explicatory remarks, and the quicker, better informed readers who, perhaps because of their former experience with *meitivi*-type queries, no longer bother to pause and think between question and answer. Reluctant to give the game away too obviously, or so it seems, the *stam* does not allow the antitraditionalism of his didactic interventions to linger long enough to really sink in. Although the problems generated by each of the *beraitot* are explicated as only an antitraditionalist could describe them, they are each immediately resolved, or rather harmonized out of existence, in ways only a dedicated traditionalist would deem fit! In other words, the framer of *Berakhot* 19b flashes his antitraditionalism repeatedly, but each time only for the briefest of moments between posing the question and suggesting an answer, never long enough for it to have a real effect. By the time each of the questions has been answered, all trace of whatever antitraditionalism was allowed to briefly make an appearance has been effectively obliterated.

7. Rav's dictum that "wherever a profanation of God's name is involved no respect is paid (even) to a teacher" is not the first mention of respect for teachers that occurs in this portion of the *Gemara*. The issue is raised on the previous page apropos a claim attributed to R. Yehoshua b. Levi to which I shall return shortly. Rav's ruling gives the impression that it is about the limits of, and the constraints upon, the courtesy one is supposed to show to one's teachers. However, and this is the third level of reading of the *sugya*, in talmudic discourse *kevod ha-rav* means more than courtesy. It refers first and foremost to respect for one's teacher's authority and especially one's teacher's *teachings*. "To dispute one's teacher" says R. Hisda, "is like disputing the Almighty," to which R. Hanina b. R. Papa adds: "To even think of doing so, is as if one was thinking of contesting the Almighty" (Bavli, *Sanhedrin* 100a). Bearing this central connotation of *kevod ha-rav* in mind, Rav's ruling acquires an additional and new aspect: if one is convinced that a profanation of God's name is in the making, Rav's seems now to be implying, one is expected no longer to pay respect to one's teacher's *teachings*. If one truly believes one's masters to be wrong and liable, as a result, to cause

a "profanation of God's name", one is *required* to set aside all considerations of honor and respect, and to speak one's mind.[3]

Read thus, Rav's principle is rendered more than the mere *outcome* of a covert antitraditionalist move. It now reads, also, perhaps even mainly, as an all but open *declaration* of antitraditionalism itself! Not only is one obliged to publicly strip if thought to be wearing mixed kinds even in the presence of one's teacher, but, Rav now seems strongly to suggest, one is required outwardly to contest one's teacher's teachings and rulings if they are genuinely thought to be inappropriate. The implication is that just as one is encouraged constantly to check all garments for mixed kinds with a view to taking action whenever thought necessary, one is likewise encouraged critically to scrutinize everything one learns with a view similarly to speaking one's mind whenever thought necessary—even in public, even to the embarrassment of one's teachers. If Rav's principle can be said to sustain such a reading (and Bavli, *Eruvin* 63a apparently presumes that it does) then it clearly implies, if not actually proclaims all the main elements of antitraditionalism.

8. The fourth and final significant aspect of this remarkable *sugya* concerns the two units of discussion that flank it on either side. Immediately preceding it is the rather strange discussion of R. Yehoshua b. Levi's claim that "In twenty-four places (it is taught that) the *bet din* excommunicates a person for acting disrespectfully towards teachers—and we learn them all in our Mishna". The fact that none of the cases mentioned in the ensuing discussion seem to have anything to do with disrespect for teachers is puzzling enough. But the strangest and most interesting aspect of the discussion, is the fact that at least two of them imply the exact opposite—namely, that individuals were excommunicated for being *too* respectful of their teachers! These are no less than the two most significant antitraditionalist tannaitic legends about sages who faced excommunication for refusing to accept their colleagues ruling *against* halakhot they received from their masters! I am refering to the story of Akavia b. Mehalal'el related in Mishna, *Eduyot* 5:6–7 and that of R. Eliezer b. Hyrqanus famously related in Bavli, *Bava Metzia*, 59a–b. Neither of them is told here in any detail. They are both briefly alluded to, assuming, so it seems, that the reader knows them well. But those acquainted with the original stories well know what Akavia and R. Eliezer were banished for! How could it be that the two most vividly

[3] Interestingly, elsewhere in the Bavli, Rav's ruling is explicitly employed to mean just that (c.f. Bavli, *Eruvin* 63a).

antitraditionalist legends in the entire talmudic corpus are recruited here as seeming confirmation of the Mishna's alleged traditionalism?! The answer, I suggest, is that here too the *stam* has ingeniously constructed a *sugya* that speaks deliberately in two distinct and easily discerned disharmonious voices that address two distinct readerships. These two voices mesh remarkably well with the two equally discordant voices that we have seen typify the *meitivi*-discussion of Rav's principle. Talk of a tannaitic policy of instituting a ban for reasons of *kevod ha-rav* introduces the kind of formal, yet hollow traditionalist superstructure, similar to the rhetoric of *meitivi*-confrontations, which those acquainted with the material can hardly fail to see as transparent. The ability to recognize this self-refuting 'Doublespeak' distinguishes the advanced from the beginners.

The final point I wish to make regarding the antitraditionalist subtext of *Berakhot* 19b concerns the culmination of the *sugya*. Here, it seems, there is no more need for subtlety or knowledge of other texts. For a brief moment, the *stam* seems to throw all caution to the wind, and to announce his meta-halakhic preferences for all to hear. He does so in the form of a dialogue between the *amoraim* R. Papa and Abaye that constitutes the boldest amoraic statement of antitraditionalism of which I am aware. The dialogue speaks for itself.

> Asked R. Papa of Abaye: In what were our predecessors different from us that miracles occurred for them though for us they do not? Could it be a question of learning? (Surely not.) In the days of R. Yehuda (for example) all they learnt (knew) was (the Order of) *Nezikin*, whereas we study (know) all six Orders—(Indeed) when R. Yehuda used to reach (the mishna in) *Uktzin* (concerning) "If a woman presses vegetables in a pot"—and some say (it was when he reached the one concerning) "Olives pressed with their leaves are ritually clean"—he would say 'I see here (a complexity of the same order as) the disputations of (my teachers) Rav and Shmuel' (but was unable to explain it himself). We (by contrast, are able to) study *Uktzin* in thirteen different ways. Yet R. Yehuda would only have to remove one of his shoes for rain to begin to fall, while we can afflict our souls and cry and cry and no one listens! He said to him: (it is not a matter of learning, rather) our predecessors jeopardized their lives for the sanctification of the Name, and we do not.

On the one hand, R. Papa's statement conveys far more than an antitraditionalist would ever need. In order to justify the adoption of a critical attitude towards the knowledge and teachings of former generations, it is quite sufficient to assume that all humans are fallible, that circumstances can change in ways that are liable to defy the most canny foresight, that even if the quality of learning of later generations is less than that of their predecessors they may still be considered dwarfs standing on the shoulders of giants. More than it may be considered a justification of antitraditionalism, R. Papa's bold assessment of the vast superiority of the knowledge and learning of his own generation in comparison to that of R. Yehuda constitutes a flat refutation of traditionalism. If it is true, it leaves no room at all for the type of commitment to former learning necessarily premised by traditionalists and seemingly presupposed in the rhetoric of standard *meitivi*-type confrontations.

On the other hand, the *stam* still succeeds in remaining ambiguous with regard to one crucial aspect. Forceful as it is, R. Papa's declaration is still not enough to prevent a traditionalist rendition of *meitivi*-type challenges completely, for the simple reason that he cautiously limits his assessment to the amoraic period. Traditionalists can still claim—and I can personally attest to the fact that many of them do!—that even if R. Papa and Abaye were right about the standard of their amoraic predecessors, it is unthinkable that they would have even considered disputing a tannaitic ruling, let alone criticizing the quality of tannas' learning. Those who failed to take note of the unnoted incongruity between Rav's ruling and the Mishna and *beraita*, who passed by the *stam*'s tedious running commentary to the five *meitivi* confrontations, who ignored the discrepancy between R. Yehoshua b. Levi's generalization and its alleged instances, will also, in all probability, read R. Papa's exchange with Abaye for its moral conclusion rather than for its epistemological premise. Despite the bold, and seemingly unequivocal message conveyed by R. Papa's assessment, the antithetical, two-tier meaning of the *sugya* as a whole is retained to the very end.

As noted at the outset, the significance of the *meitivi* confrontations of *Berakhot* 19b is enormous precisely because, *qua meitivi* confrontations, they are wholly and totally typical. The five tannaitic challenges to Rav's ruling and their subsequent resolutions are so commonplace that, when studied, they are normally skimmed through with ease. They are not a parody of their kind in any obvious sense. The ways in which they are resolved by the *stam* do not appear to ridicule such solutions. On the contrary, the force and subtlety of the *stam*'s ironical exposition of their real meaning owes much to the fact that, of themselves, they are entirely credible and wholly representative of the Bavli's vast stock of similar moves. This is why I suggest that *Berakhot* 19b should be seen as a paradigm rather than a parody; as an extraordinarily well-crafted *sugya* that contrives to explain rather than satirize the class of *meitivi*-type, transgenerational negotiations. Viewed thus, *Berakhot* 19b is seen as an instructive, explanatory effort on behalf of one of the framers of the Bavli's many other *sugyot* of its kind, rather than that of an antagonistic critic. It is, I urge, the work of an antitraditionalist doing his best to *disclose* the Bavli's antitraditionalist project, rather than that of an antitraditionalist aspiring to ridicule a traditionalist one.

A MODEST PROPOSAL: TOWARD A RELIGIOUS POLITICS OF EPISTEMIC HUMILITY

Menachem Fisch

Preliminaries

Before applying ourselves to the task set to us by the editor of this volume—that of thinking through the prospects of developing a position of toleration that is articulable from within the assumptions of a revealed religion—it is wise, I think, to first define what we mean by tolerance, and what we think toleration per se is capable of providing. My reason for raising this question is not to bog us down with preliminaries, but, rather, to propose raising the ante. To put it simply, from the perspective from which I am writing—philosophical and religious—tolerance, desirable as it certainly is, will simply not suffice.

A few preliminaries. First, I shall be talking about tolerance in the strong sense of undertaking to put up with those with whom we disagree or find objectionable, to be willing to endure and to quietly suffer the discomfort of their presence—as opposed merely to the toleration of the foreign or the unfamiliar. It is the objectionable, rather than the simply different other on whom we should concentrate. To articulate a position of tolerance from within the assumptions of a revealed religion is, hence, minimally to find religiously meaningful ways to justify systematically turning a blind eye on systems of belief, forms of life and modes of conduct considered heretical or sinful by the religion in question.[1]

* This article was first published in: Menachem Fisch, "A Modest Proposal: Toward a Religious Politics of Epistemic Humility," *Journal of Human Rights* 2, no. 1 (March 2003), pp. 49–64. Reprinted with permission of Taylor & Francis LLC (http://www.tandfonline.com).

[1] The analytical framework of this paper owes much to the work of Israeli philosopher Avi Sagi. First, to the sharp distinction between tolerance and pluralism set forth (somewhat more stringently than here required) in his "The Jewish Religion: Tolerance and the Possibility of Pluralism" (Hebrew), *Iyyun*, 47 (1995), pp. 175–200, and developed further in *Elu va-Elu: A Study on the Meaning of Halakhic Discourse*, (Hebrew) Hakibbutz Hameuchad, 1996, especially pp. 194–7 and "Religious Pluralism Assessed", *Sophia*, 38 (2), 1999, pp. 93–115. (The notion of pluralism I adopt below partly coincides with a variant of the "weak" or "moderate" of the three forms of pluralism he discusses—see ftn. 17 below.) And second, to his "On Tensions Between the Orthodox and the Secular: Rights versus

Second, to pose the problem even more pointedly, I suggest we avoid entirely the option of making more room for the other by lowering religious standards. In encountering modernity in its various forms and phases all religions have modified their notions of sinfulness and heresy in certain respects and to certain degrees. This is undoubtedly as fascinating a process as it is complex, but it is not our concern at present. The problem of tolerance with respect to revealed religions—religions with a clear and binding sense of inclusion and exclusion, right and wrong, proper and improper—is *not* that of them finding ways of *justifying* heretical and sinful conduct, but of them finding ways to justify *refraining from taking action* against such conduct *despite* it being heretical or sinful. What we are here to explore from the point of view of revealed religion, are the prospects not of rethinking sin, but of rethinking *reaction to* sin.

Third, the problem does not arise for all forms of life and modes of conduct held sinful by the religions in question. Those held in similar contempt by the rest of society, even if for other reasons, may still pose a problem to some, but cease to do so from the point of view of the concerns of the present volume. I take it that we are not seeking ways to render the religions more tolerant than the rest of the world. The problem arises not for the attitudes of revealed religions to sin in general, but to such forms of sinful conduct that are tolerated outside their communities. One can assume that these will normally include most violations of ritualistic religious norms—those to do with taboos and exclusions concerning foods, rites, prayer, and to an extent people. They will minimally exclude most acts of violence against members of other communities and violations of their private or communal property.

Neither, I think, should we be concerned in this particular context with intolerance expressed by religious communities towards their own heretic or sinful members—at least not directly. Assuming that such communities are autonomous, at least to some degree, and setting aside the very important question of determining how illiberal a liberal multi-cultural society can allow its communities to be with respect to their own members, I understand our interest to lie with the capacity of the religious to

---

Identity Discourses", in N. Ilan (ed.) *A Good Eye: Dialogue and Polemic in Jewish Culture*, Hakibbutz Hameuchad, 1999, pp. 408–30, and forthcoming "Israeli Law and Society: Rights versus Identity Discourses", *Mehkarei Mishpat* (2000), for the claim, argued for below, that liberal rights discourse is capable of sustaining only tolerance, but not pluralism. The present paper can be viewed as a friendly and critical engagement of his work with a view to raising and solving a problem he has not addressed by means of a position of which he is critical.

tolerate the heretical and sinful conduct of communities and individuals other than their own. The problem of religious toleration thus arises with respect to beliefs, actions, individuals and forms of life that are tolerated, and may even be valued beyond the confines of the religious community within which, and in the eyes of which they are deemed intolerable.

The Problem of Sovereignty—A Jewish Perspective

That being said, the first thing to notice is that from the point of view of revealed religion there are two, rather than one problem to consider: that of religious communities dealing with the objectionable behavior of *neighboring* communities and individuals, and the very different problem of religious individuals dealing with such behavior in the capacity of members of the governing body or ruling class responsible for the communities or individuals in question. For the former, tolerance, in the manner defined above, is all that is needed; for the latter, I shall argue, tolerance is not enough. And since it is the latter problem—that of religious sovereignty, as opposed to that of mere peaceful coexistence—that, in my view, presents the gravest challenge to observant members at least of my own religion, I shall devote the main bulk of my comments to explaining why this is so, and to exploring a possible way out from which, I shall suggest further, a wider lesson might be learned. But first a word about the problem and prospects of peaceful coexistence from a Jewish point of view.

Most existing ideas of tolerance, states Adam Seligman in his introductory remarks, come grounded either in considerations of Real Politik, or in those to do with liberal assumptions on the nature of humankind. We tolerate other people's beliefs according to the former because it is politically worth our while. We do so according to the latter, liberal, option, either because one's beliefs are regarded one's private business, or because individual autonomy is held sacred. In Judaism, belief per se poses no real problem. For Judaism, the need for tolerance asserts itself almost exclusively in the realm of the practical, the realm of improper conduct. With regard to sinful conduct (e.g. public 'desecration' of the Sabbath, public display by Jews of foods prohibited during the feast of Passover, open single-sex communions, idol-worship) only the former is applicable. Jewish law—halakha—contains, as it stands, nothing comparable to the sanctioning of individual autonomy in modern liberal thought. In halakhic discourse, freedom of choice, and the essentially voluntary basis of halakhic obedience that it entails, are of crucial theological importance. But it is the *possibility*, rather than a valued *right*, to chose evil that grounds ideas of personal halakhic accountability. For halakhic Judaism as it stands the only viable option with regard to the toleration of sinful conduct is Seligman's first, although 'Real Politik' is perhaps too harsh a term.[2] The halakhic principles used to justify slackening halakhic standards in order to avoid unnecessary social or political tension are: *mipnei darkei shalom* and *mishum eiva*—that is to say, for the sake of peace, or for fear of hostility, which, in the last analysis, boil down to rather crude, ad hoc, and wholly practical versions of Kant's categorical imperative—not: do not do to others what you would prefer others not to do to you, but rather: avoid from doing to others what the law would have otherwise required you to do to them for fear of them failing to appreciate your noble intentions, and retaliating by doing the same or worse to you. Such reasoning, as we shall see shortly, is not as morally coarse as it may sound. It is pragmatic for sure, though not in the worst sense of the term. For it presupposes a real acknowledgement of, and willingness to grudgingly accommodate, ethical diversity. It implies a real realization, to which many religious systems have proven themselves impervious, that what may be right according to my system may look exceedingly wrong from the perspective of another. It is sincere, and, most importantly, it works quite well.

Most Religions, Judaism in particular, discriminate in two different and largely independent ways. They discriminate, on the one hand, between prohibited and permitted conduct, and discriminate, on the other, between people: between members of the faith and those who are not, or even between members of a faith, any God-fearing faith, and those who are not.[3] (I am ignoring here such principles of exclusion that are applied within the community—e.g., between young and old, male and female, married and single, etc.) In a variety of situations, Jewish law discriminates against gentiles qua

2 See also Sagi, "Jewish Religion", p. 198.

3 Such, for instance, was the way in which the thirteenth-century Provencian halakhist and commentator on the Talmud, Menahem Ha-Meiri, intriguingly redefined idolatry. For all halakhic purposes, he rules, an idolater is not a worshiper of other gods, but a person "unbound by the practices of religion"—*any* God-fearing religion, that is. According to Meiri all forms of Islam and Christianity are thus ruled non-idolatrous. An important consequence of his bold move is a dramatic reduction, within the boundaries of religious restriction, of most forms of halakhical exclusion. Unfortunately his monumental work, *Beth Ha-Behira*, was lost during the crucial years in which the great halakhic systems of Maimonides, Joseph Karo and Moshe Isserlish became the norm. For clear statements of his position see *Bet Ha-Behira* to Bavli, Avoda Zara 15b, Bava Kama 112b and especially Bava Metzia 59a where he goes so far as to categorize a people bound by the ways of religion as "a people that is with you in Torah and commandments"—*am she-itkha be-Torah u-be-mitzvot*!

gentiles—sinful or not. Hence one may speak here of two separate forms of religious toleration: a willingness, in certain circumstances, to put up with forms of conduct otherwise deemed objectionable, and a willingness, in certain circumstances, to include those who would otherwise be excluded. Both are easily accommodated by the pragmatism of arguing "for the sake of peace", and, on occasion supplement each other. "In a city populated by Jews and gentiles, Jewish and gentile administrators are appointed", states the Jerusalem Talmud, "taxes are collected [equally] from Jew and gentile, [from which] the needs of the Jewish and gentile poor are [jointly] supplied." One does not discriminate between the Jewish and gentile sick, the Talmud goes on to rule, one attends equally to the dead of each community, and pays one's respects equally to their respective mourners—all in the interest of peace.[4] Concern "for the sake of peace" thus allows one to passively ignore sinful conduct and requires one actively to attend to the immediate needs of those engaging it. (As the Yerushalmi makes quite clear, the gentiles in question are active idolaters.) For the sake of peace sin can be ignored, and the needs of sinners attended to. One is supposed to attend, however, only to needs they share with the rest of humankind. They are not to be actively assisted *in their sinful ways*. There is no obligation to supply them with places and implements of worship, for example, in fact it is downright forbidden. Jewish law, one could say, fails to recognize a sinner's right to his or her sinful way of life.

When Jews find themselves sharing political space with gentiles, these early sources rule, they are indiscriminately obliged to share the responsibility for burying the dead, attending to the poor and the sick and condoling the mourners of the two communities (stopping short of heeding to their cultural and religious needs insofar as these violate halakha). But in post-talmudic halakhic texts, the extent even of this supposedly most humane of obligations is debated—a debate that has been made worryingly relevant to modern Israel. At one point in his monumental halakhic codex,[5] Maimonides explains that we are obliged to do so because it is our moral duty "The Lord is Good to all: and his tender mercies are over all his works" (Psalms 145: 9). Elsewhere, however, he insists, that the obligation applies only when and where Israel is "suppressed by the hand" of the gentiles. However, when Israel has the upper hand and is predominate in its land, not only are we no longer obliged to attend the needs of gentiles

[4] Talmud of the Land of Israel—Yerushalmi—*Gittin*, v, 47c, citing and modifying Tosefta, 3: 13–14.

[5] Maimonides, *Mishne Torah*, hilkhot melakhim 10: 12.

in residence, but are forbidden by halakha to grant them temporary residence or even the right to enter the land. "The foregoing rules [of conduct toward gentiles for the sake of peace]," he writes,

> apply to the time when the people of Israel live exiled among the nations, or when the gentile's power is predominant. But when Israel is predominant over the nations of the world, we are forbidden to permit a gentile who is an idolater to dwell among us. He must not enter our land, even as a temporary resident; or even as a traveler, journeying with merchandise from place to place, till he has undertaken to keep the seven Noachian Laws.[6] (…) For those who accept the Seven Laws are regarded Resident Proselytes.[7]

However, Maimonides continues, Resident Proselytes are only accepted during such times in which the law of the Jubilee is in force (the fundamental fifty-year cycle (Lev. 25: 10–12) that Maimonides himself maintains has now been irretrievably lost).[8] Otherwise only full proselytes—full converts to Judaism—are accepted. Tolerance toward the gentiles with whom we share social and political space, is limited, according to Maimonides, to exilic conditions. Once fully sovereign in their land, halakha no longer allows Jews to tolerate even the mere existence of gentiles among them, let alone attend to their needy. Importantly, Maimonides's ruling applies to all gentiles as such, and is not limited to idolaters.

Needless to say, not all agree with Maimonides on this point, and among those who, in principle, do accept his ruling, fewer still regard the conditions of modern Israel as meeting the requirements of Maimonidean sovereignty. Some interpret him to mean that only idolaters should not be tolerated by a sovereign Israel, but contest his views on who is an idolater. Contrary to Maimonides, many follow Menahem Ha-Meiri in denying that Christianity constitutes a form of idolatry at all. (All agree, including Maimonides, that Islam is not a form of idolatry.) Since the "gentiles of our time" are not idolaters, they argue, Maimonides' ruling no longer applies. (He himself, however, drew the line, as we have seen, not between idolaters

---

6 These are laws supposed by the rabbis to have been binding upon all of mankind prior to Sinai, and upon non-Jews thereafter. There is noticeable disagreement in the talmudic texts regarding the precise list of Noachian laws. Basing their exegesis on *Genesis* 2:16 and 9:4 some sources list the following: (1) not to worship idols; (2) not to blaspheme the name of God; (3) to establish courts of Justice; (4) not to kill; (5) not to commit adultery; (6) not to rob; (7) not to eat the flesh cut from a living animal (*Genesis Rabba* 16:9, 24:5). Elsewhere (Talmud of Babylonia—Bavli—*Sanhedrin* 56b) blasphemy and the establishment of courts are dropped in favor of the emasculation of animals and the pairing of animals of different species. Elsewhere still (Bavli, *Hulin* 92a) the list is much extended.

7 *Hilkhot Avodat Kokhavim* 10:5–6.

8 *Hilkhot Shmita ve-Yovel* 10:3–7.

and the rest of humanity, but (given the fact that Resident Proselytes no longer exist) between Jews and the rest of mankind. From this perspective the definition of idolatry is irrelevant!)

Rather than contest his notion of idolatry or depiction of Israel's desired citizenry, others still have challenged the realizability of his very notion of sovereignty. No democratic government can, in principle, be ever fully sovereign in Maimonides' sense of the term, argued the late Chief Rabbi of Tel-Aviv, R. Haim David Halevi, for no such government is at liberty to do as it pleases. Democracy, by definition, is a form of government to which Maimonides's very concept of predominance is inapplicable. Hence, urges R. Halevi, even if one commits oneself to abiding by Maimonides's rulings on these matters (which he himself, in following Ha-Meiri, does not), they remain irrelevant to modern Israel.[9] The point, however, is not the extent of disagreement among halakhists with regard to the passive toleration of sin, so much as the impressive lack of it with regard to actively assisting the sinner. On this important issue all appear to agree. Idolaters or not, as noted above, halakha, firmly forbids attending to the *religious* needs of non-Jews (let alone actively supporting the anti-halakhic lifestyle of irreligious Jews!). R. Menahem Ha-Meiri, whose rulings regarding toleration are the most radically lenient, summarizes the point concisely in his commentary to Bavli, *Avoda Zara* 11b. Citing the Yerushalmi's ruling mentioned above, obligating the governors of mixed cities to tax Jews and gentiles indiscriminately, and to attend equally to their poor, sick and the dead, all for the sake of peace, he goes on to caution that "their religious feasts, on the other hand, are not allowed, even for the sake of peace, as is any action suspect of idolatry." Concern for peace has its limits: passive toleration, yes; attending to a sinners' human needs, yes; it stops, however, at actively supporting sinful conduct. On this, I believe, all agree.

There exist ample halakhic resources for toleration in the narrow sense of the term. In principle, at least, observant Jewish communities encounter no problem refraining from taking the offensive action otherwise required of them by halakha, whenever there is even the slightest chance of retaliation. The pragmatic meta-ethics that backs overriding halakhic prohibition by means of principles like "for the sake of peace" may not always be to everyone's taste (e.g. when looking for ways around the halakhic prohibition against desecrating the Sabbath to save the life of a non-Jew), but they

---

[9] R. Haim D. Halevi, "*Darkhei Shalom* in Relations between Jews and Non-Jews", *T'humin* 9 (1988) pp. 71–81.

do supply all that is needed for halakhically committed Jews to live peacefully among the sinful, the heretic and the gentile. They do not suffice, however, to *rule* the sinful—as the example of the mixed city clearly implies. To this end, I submit, mere toleration along with the metahalakhic principles that ensure it are not nearly enough.

A simple example will suffice to make the point. Some years ago Rabbi Eliyahu Bakshi-Doron, Sephardic Chief Rabbi of Israel, was asked by an observant Jewish travel agent whether halakha permitted him to cater for non-Jewish pilgrims to the Holy Land, whose numbers, he believed at the time, were expected to rise dramatically in the wake of the Oslo agreements. I shall not burden you with the details of the Chief Rabbi's response, which is relatively short and decisive. He clearly sides with Maimonides's rather than with Ha-Meiri's ruling concerning Christians. "Christians believe in a shared Deity and they worship images", he submits,

> Hence, their faith contains an element of idolatry. He who promotes religious Christian tourism, is, in effect, promoting their rituals of idolatrous of worship (…) To assist them in actual idolatrous worship, by attending to their transportation to 'that forbidden place', and helping them gain entry to where those rituals are performed, seems to be not only a transgression of "Thou shalt not put a stumbling block before the blind" (Lev. 19:14), but a denial of our religious obligation to uproot all idolatry from our land, as stated by Maimonides (…) It is true that considerations of *eiva* and *darkhei shalom* prevent us from performing this halakhic duty [sic!], but it is certainly prohibited to help them to worship in a place, and thereby rendering it a house of idolatrous worship. It appears, therefore, that one should refrain from encouraging Christian religious tourists or assisting them in visiting and performing idolatry in the places sacred to them. However, it is of no concern to the travel agent if regular tourists [as opposed to pilgrims] decide, in the course of their visit, to take the opportunity to pray and perform idolatry.

Although one is not required to actively hinder Christians making their way to their places of worship, halakha clearly prohibits actively directly assisting them in doing so, let alone profiting from it in any way. "So" the rabbi concludes,

> Let the person who raised the question be rewarded twice, once for his obviously genuine concern for peace, and again for concentrating hereafter on encouraging and helping bring many, many Jews to The Holy Land.

With regard to private travel agencies Chief Rabbi Bakshi Doron's ruling poses no real problem. It sounds awful, and terribly impolitically correct, but, in principle, can be lived with. Every religion has certain areas of seclusion from which non-believers are barred. Such being the halakha (and as far as I can tell, if Maimonides is to be followed on this, the Chief Rabbi's reasoning is halakhically sound), one can easily imagine separate travel agencies catering for orthodox communities to the ritual exclusion of others, just as they are catered for by special kosher restaurants, bakeries and wineries. Paradoxically, such exclusionary institutions function as social insulators who, by accentuating the community's particularity, render the resolve to live and let live all the more easier to justify, and certainly easier to maintain. But imagine for a moment that the person seeking the Rabbi's advice had not been a private entrepreneur but an orthodox member of the Israeli Cabinet.

The role of government in multicultural democracies differs decisively from that of communal leadership. While the latters' job, even when acting as representative of their home communities or cultures, is primarily to look out for their own, the former, regardless of personal conviction, assumes responsibility for the safekeeping and wellbeing of all. The Israeli Ministers of Education, Religious Affairs, The Interior, Tourism, etc., will not have performed their governmental duties if all they can boast is a willingness to turn a tolerating blind eye on forms of conduct their religion(s) consider sinful. It is their job, in their designated areas of responsibility, to actively ensure the security, the welfare and the capacity of each and every community to flourish according to its custom and conviction (insofar as, in doing so, none harm others, or prevent other communities from exercising their own, identical rights, of course). If halakha firmly forbids Jewish travel agents to actively assist or profit from Muslim and Christian pilgrims,[10] it similarly forbids Jewish governmental officials to do so. But to volunteer such active assistance is the very job of the Ministries of Religious Affairs, Finance, Tourism, and Education, to name but the most prominent. Not only are these ministries expected, for example, to assist Israel's Muslim and Christian communities in planning and funding their annual school trips to the places holy to their faiths, but, as a matter of course, are held responsible for them being able to do so most efficiently and effectively. It is their job, in other words, not merely to safeguard the religious freedom of those deemed idolaters by many halakhists, but to assist them any way they can. And the same goes for the gravely sinful lifestyles of unorthodox Jews. The very same ministries are responsible for the ongoing activities of

---

[10] Rabbi Bakshi Doron makes clear that his ruling equally applies to Muslims despite the fact that, following Maimonides, Islam is not considered idolatry. His reasoning there, however, has to do with their religious claims to the Temple Mount, and the halakhic problems to which their very presence there gives rise.

the various Israeli youth movements—some of which are decidedly, even ideologically irreligious. Most of their out-of-town activity—field trips, camping outings, hikes—are scheduled for the days held most holy by halakha. And, again, personal convictions notwithstanding, it is the Minister's job not only to assist and fund such blatant and public desecration of the Sabbath and High Holidays, but to study it year by year, learn from past mistakes, and render it all the more enjoyable, efficient and effective for those involved! This is a far cry from undertaking grudgingly and silently to passively ignore sin for the sake of peace.

Needless to say, *because* of its built-in sensitivity to the right of all people to what *they* define as their own culture, lifestyle and religion, modern multicultural democracy is by far the form of host government best suited and most desired by orthodox Jewish communities. The autonomy granted by such states provides, prescribes, ensures and protects the cultural, educational and religious space needed for religious communities to live their lives as they see fit.[11] And yet, as things now stand halakhically, this is, at the same time, the form of government most awkward for observant Jews to adopt in order to rule others. What is required of halakhically committed members of such a government is much more than tolerance. In order to justify going out of their way to actively protect and assist what they consider to be gravely sinful forms of conduct, the halakhically committed Jew requires some form of religiously justified pluralism. But how is it possible to square pluralism with an exclusivist religious viewpoint advocating "a system of thought and action that offers a vision of ultimate Good based on transcendental truths", as Seligman aptly puts it? On what grounds can halakhically committed Jews view forms of conduct that are decidedly rejected within their religious communities as genuinely worthy of promotion outside them? As we have seen, none of the three philosophies of tolerance outlined by Seligman is capable of grounding the sort of double-standard, or, to borrow Yehuda Elkana's term, two-tier thinking that seems to be required here.

In remainder of this paper I shall propose a possible way out. My exposition will be sketchy, and hesitant. The theory of toleration I shall be opting for, while not new to political philosophy, has not gained the currency its

---

[11] On the question of one's right to the culture of one's choice in contemporary liberal philosophy see: W. Kymlicka, *Liberalism, Community and Culture*, Clarendon Press, Oxford, 1989, Ch. 8; A. Margalit and M. Halbertal, "Liberalism and the Right to Culture" (Hebrew), in M. Mautner, A. Sagi and R. Shamir (eds.) *Multiculturalism in a Democratic and Jewish State*, Ramot, Tel-Aviv, 1998, pp. 93–105; A. Sagi, "Israeli Law and Society".

author, Sir Karl Popper, had hoped for. It remains, by any standard, a minority voice—unjustly so in my opinion, but a minority voice anyway. Hence, it is a position, for which a case needs to be made independently of its application to our present concerns. Needless to say, even if such an option is, in principle, deemed viable, its applicability to the specific context of religious commitment, requires separate justification. Finally, if and when that too is accomplished, it will still remain to be shown that halakhic Judaism, the specific perspective from which I am writing, possesses resources sufficient for accommodating such an approach. To set forth such a three-stage argument in acceptable detail would require at least a book. What follows, then, will not be a summary of work accomplished so much as an initial report on work still very much in the making. For lack of space, and in order to remain as fully focused as possible on the problem of *religious* toleration to which this volume is addressed, I shall skip the first stage of the argument completely, and assume that the general approach advocated is, at least from the point of view of political philosophy proper, a viable option. Thus equipped, I shall briefly explain why I think it is an attractive one for the task at hand, and go on to give my reasons for maintaining that halakhic Judaism is capable, in principle, of accommodating such an approach.

A Different Approach: Learning From the Other

Though a far cry from Real Politik, the approach to tolerance I have in mind does not require the apriori endorsement of what he dubs "liberal assumptions on the nature of man". It is not grounded in 'rights talk' to use Bruce Ackerman's coinage.[12] I am referring to the set of presuppositions on which Karl Popper bases the theory of toleration outlined in his *Open Society and its Enemies*.[13] These primarily are *not* to do with a moral duty to defend or

---

[12] Unlike Sagi, I use the terms 'rights talk' and 'rights discourse', as does Ackerman, to denote the philosophical discussion of rights as the basis of liberal theory. Sagi, in contrast, uses it to denote the practical negotiation of rights within society, typically arbitrated by the law. Cf. B. Ackerman, *Social Justice in the Liberal State*, Yale University Press, New Haven and London, 1980, p. 5. Ackerman is a case in point in likewise refusing to view rights as a primordial apriori given. For him too the notion of rights is a derivative of what he calls the principle Rationality. Here, however, is where we differ significantly. See next ftn.

[13] K. R. Popper, *The Open Society and its Enemies*, 2 Vol., Routledge & Keagan Paul, London, fifth edn. (revised), 1974. The book is devoted almost entirely to critical evaluations of the social and political philosophies of Plato, Marx and Hegel. Popper's own views are developed mostly in extended footnotes. See in particular Vol. I, p. 235 ftn. 6, pp. 237–9 ftn. 18, pp. 265–6 ftn. 5 and pp. 287–91 ftns. 6 and 7. For a brief yet useful overview of his

respect individual autonomy, or with a moral obligation to keep out of one's neighbor's hair. Popperian tolerance, if I may call it so, *entails* such liberal presuppositions of individual autonomy (if, as we shall see, interestingly constrained), while *deriving* ultimately from epistemological, rather than moral or metaphysical arguments. Belief systems and lifestyles other than their own are given voice and allocated space by members of the open society primarily because they realize that a plurality of engaging, especially conflicting voices provides the best setting for most effectively articulating and developing one's own. The close presence of challenging, even disturbing and dissenting voices is welcomed, because the chances of one's awakening to error and oversight are exceedingly greater when faced with a variety of real and serious alternatives.[14]

The obvious advantage of grounding ideas of tolerance in this way is that the arguments for them involve far less by way of ontological, ideological, psychological, moral and metaphysical presupposition—all of which are fiercely debated. Most importantly it sidesteps the entire issue of fundamental rights, an idea halakhic Judaism for one has great difficulty accommodating as an apriori given. The acknowledgement of humankind's profound fallibilism especially in the public sphere, of the fundamental timeboundedness and context-dependence of all social planning, coupled to the almost built-in instability of those very contexts, premises little more than that life, especially social life, is exceedingly complex and unpredictable, and that the chances of us making a mess of things are enormous—with this few would disagree. The upshot is an epistemic hesitance, modesty and openness, an awareness that we are poor critics of our own plans and strategies, that things are best done piecemeal, and that other viewpoints are not merely to be tolerated, but to be highly valued, not for

---

position see his "Toleration and Intellectual Responsibility", in his *In Search of a Better World*, Routledge, London, 1984, pp. 188–203.

14 For Bruce Ackerman, people's possession, or right to power rests on their ability to provide reasons that explains why they are entitled to the resource in question (*Social Justice in the Liberal State*, p. 4). "Rights", he argues, thus "have a reality only *after* people confront the fact of scarcity and begin to argue its normative implications." (p. 5) One employs rational argument, according to this view, in order to *resist* granting others a share of one's political and cultural space. Others are granted their share, exactly when and where rationality fails those in power. Following Popper's lead, the position I argue for also grounds rights in rationality, but quite differently. We grant others a right to power not because we are incapable of explaining why we shouldn't, but because we have a profoundly rational need for having them around, owing to their ability, by virtue of their very otherness, to question our own political and cultural choices.

their own intrinsic worth so much as for their capacity to challenge our own by helping highlight its flaws and shortcomings.

Let me remark in passing that the Popperian option offered here in order to counter the dogmatism of many religious systems, is equally effective in countering other forms of uncritical political discourse. By denying the self-correcting force of argument and criticism, modern (or rather post-modern) relativism denies multi-culturalism any epistemic value. According to Richard Rorty and his ilk, language is no longer allowed meaningfully to refer or to express, and reason, no longer to act as an interpersonal court of appeal.[15] Humans are described as accounting for their world and for themselves within it by means of 'vocabularies' of metaphors which they acquire, inherit or construct, without the benefit of rational procedures to criticize them, compare them or pit them against each other. As a result, the liberal societies envisaged by these thinkers lose their value as non-violent, non-manipulating forums for the critical exchange and amelioration of ideas. As Rorty's candid *Contingency, Irony, Solidarity* amply proves, the very idea of such a society is thrown into sharp relief with the extremely individualistic, self-centered, and unchecked egocentrism that is all but entailed by their flight from reason. (His recent *Achieving Our Country*[16] shows, in my opinion, the beginnings of an awakening to the dire political consequences of his earlier views, but that is the topic for an entirely different paper.) The Popperian option thus serves as a Janus-faced alternative jointly opposed to both uncritical extremes: uncritical dogmatism on the one hand, and the equally uncritical position of absolute relativism on the other.[17] For obvious reasons the present paper will concentrate on the former.

---

[15] Rorty's views on language and knowledge and their bearing on political theory acquire their fullest development to date in his *Contingency, Irony, Solidarity*, Cambridge University Press, Cambridge, 1989.

[16] R. Rorty, *Achieving Our Country: Leftist Thought in Twentieth-Century America*, Harvard University Press, Cambridge Ma., 1998.

[17] This is the point at which my position diverges most significantly from Sagi's, whose recent work has come to favor an explicitly Rortian form of cultural and religious plurality. He advocates for religions a dynamic 'expressive pluralism' recognizing "the basic fact that an individual's attachment to a particular religion is not the result of a critical examination of this religion" ("Religious Pluralism", p. 106). In the context of discussing religious diversity (in which he correctly includes "the deniers of religion", *ibid.* p. 108) the Popperian option (he terms 'weak pluralism'—*e.g. Elu va-Elu*, pp. 192–3), discussed by him in other contexts (most significantly within the boundaries of halakhic discourse, on which more below), drops out of the discussion completely when talking of inter-religious pluralism, and is past over all too briefly in favor of a similar position in his more general treatment of multi-culturalism. For explicit allusions to Rorty see "The Orthodox and the Secular" pp. 424, 426.

The great philosophical achievement of Popper and his school lies in their success in premising a genuine and perpetual skepticism (of the retail, rather than wholesale variety)[18] without allowing it to become destructive. On the contrary. Despite their insistence that the recognition of general truth is, in principle, beyond human ken, and that even the physical sciences, let alone social and political science, are wholly incapable of ever confirming their theories, they succeeded in formulating one of the most optimistic and constructive visions of science and society framed in this century. This they achieved by not defining progress (as opposed to mere change), even scientific progress, in cumulative terms, but in terms of a perpetual process of problem-seeking and -solving. Scientific progress, according to Popper, is not a matter of ever-increasing our stock of reliable, well-confirmed theories precisely because reliability cannot be humanly vouched for. Rather, one should view the growth of knowledge not in terms of anything 'growing' or increasing, but in terms similar to those used in industry and evolution theory: of the improvement, perfection, amelioration, adjustment of its explanatory systems and models. The key to scientific accomplishment is to be found not in the accumulation of evidence supportive of our efforts, but in our ability to keenly expose their shortcomings and propose better suited alternatives. I have shown elsewhere how Popper's vision of the growth and rationality of science, can be systematically generalized to include all manner of human endeavor by speaking in general of acting upon any goal-directed system.[19] A system is said to have been ameliorated or improved exactly if to the best of our knowledge problems it was known to harbor have been, at least, partly solved. To act rationally on a system, on any system, is to adopt a deliberately critical attitude towards it, to be constantly wary of its performance regardless of its apparent successes, and to be willing to modify it if it is ever found wanting. Once rationality is identified, as it is here, with perpetual self-doubt, with knowing one's limits, with systematic self-criticism and a constant and creative search for better-suited alternatives (rather than with focusing one's intelligence on justifying one's position a lá Ackerman, and defending

[18] See Michael Williams, *Unnatural Doubts: Epistemological Realism and the Basis of Skepticism*, Princeton University Press, Princeton, 1996), esp. Ch. 1.2. Although this is one of the most thorough discussions of skepticism and its modern responses, here too the Popperian option is hardly touched upon. For further details with reference to the distinction between casting total and piecemeal doubt, see M. Fisch, *Rational Rabbis: Science and Talmudic Culture*, Indiana University Press, Bloomington, 1997, pp. 5ff.

[19] M. Fisch, "Towards a Rational Theory of Progress", *Synthese*, 99, pp. 277–304, 1994; *Rational Rabbis*, pp. 3–39, and above pp. 35f.

the system in hand), it is easy to see how a healthy self-centered critical pluralism follows suit.[20] Other opinions are not merely tolerated. They are sought after and valued for their very otherness; for the way in which they challenge and exercise our own, help expose our mistakes and weaknesses and present, by virtue of their otherness, serious, often unthought, and certainly untried alternatives.

Grounding political philosophy in this manner does not entail *denying* the idea that people have a fundamental right to the belief system or inoffensive form of conduct of their choice. The liberal argument from rights and the argument from epistemic humility proposed here are independent of one another though by means mutually exclusive. The former, however, presupposes much, and, taken alone, promises very little. Even if we ignore the problems currently exercising liberal thought—such as the much-discussed question of how liberal a liberal government should be toward illiberal attitudes and practices within its autonomous communities—the argument from rights and the liberal theory it entails provides little incentive for building and maintaining a society transcending the boundaries of one's own community. A willingness to grant all people the right to the belief systems of their choice offers, in and of itself, no reason at all for wanting to share a multicultural framework with them to start with. It entails an obligation on behalf of governments to attend to all (non-violent) communities evenhandedly, but it is their *right* to their cultural choices that is valued not the cultural choices themselves; it affords people the rights of speech, belief and (non-violent) practice, but gives no reason for anyone to listen or engage. If the only principle doing social and political work is the acknowledgement of rights (by a leadership driven by "the anxiety of influence", the horror of inoriginality, as Rorty, following Harold Bloom, aptly puts it),[21] it comes at no surprise that Rorty encounters such enormous difficulties promoting "a genuine sense of human solidarity". But one doesn't have to buy into Rorty's entire critique of philosophical theorizing in order to realize this. Liberal rights talk provides sufficient grounds for unresentfully putting up with cultural and ethical diversity once it has developed, but gives no reason for desiring it in the first place. No intrinsic value is attached to otherness. To repeat, acknowledging a people's right to their beliefs and cultural choices entails an obligation to

[20] The term "critical pluralism" was introduced by Popper to distinguish it from relativist renderings of the term 'pluralism'. See Popper, "Toleration", p. 191, and ftn. 17 above.

[21] Rorty, *Contingency, Irony, Solidarity*, Chapter 2 refering to Harold Bloom, *The Anxiety of Influence*, Oxford University Press, Oxford, 1973.

tolerate their holding to them, but not to take any interest in their content. Of itself, rights talk attaches no intrinsic value to beliefs and choices different from one's own.[22]

Premising no more than genuine epistemic humility, the Popperian option does. The pluralism to which it give rise is not a mere by-product. It lies at the very heart of the enterprise. The combination it premises of robust yet constructive skepticism promotes a cheerful acceptance and genuine curiosity towards other cultural choices. One wants to surround oneself with people who think and live differently in order to enrich one's own beliefs, moral standards and cultural choices. On this inherently pluralistic model the other is not merely tolerated, he is welcomed and valued for his otherness. Much more is achieved this way although much less is premised. Pluralism is achieved without fundamental rights having to be even premised, let alone negotiated. This in itself renders it an exceptionally attractive option for religious systems whose capacity for accommodating a rights-based liberalism is severely limited. Arguments from rights compel us to take seriously a person's freedom of choice, but, of themselves, attach no intrinsic value to the choices themselves. Arguments from epistemic modesty, by contrast, attach great value to the outcome of their choosing, but, of themselves, care little about their liberty to do so.

Furthermore, and perhaps most importantly for the particular problem I am raising, by allowing one to set aside the entire rights issue, the option I am proposing renders quite natural the sort of two-tier thinking we found so baffling in the previous section. Paradoxically enough, the seemingly impossible double-standard required is not merely justified, but is actually encouraged within such a critical plurality. To recall, the problem of religious sovereignty is that of religiously justifying governmental executives working actively to ensure the security and prosperity of forms of conduct considered sinful in the eyes of their own religion. From the critical perspective of the type of pluralism I am proposing it is far less paradoxical than it sounds to be able to view forms of conduct that are firmly condemned within one's own community, as genuinely worthy of active support outside it. The Open Society ideally comprises communities guided morally, spiritually and politically by epistemically humble learners, rather than systematic agnostics. They themselves realize that the learning process is greatly enhanced by them being made aware of workable

[22] For forceful presentations of this argument see Sagi "The Orthodox and the Secular" and "Israel Law and Society".

alternatives. If one is genuinely interested in effectively testing and improving one's own system, being able to compare it to other, real and seriously thought-through alternatives, is always an advantage. If, on the other hand, the communal leadership considers its main task to be not to test, but to sustain, protect or perpetuate their own system, the very proximity of challenging alternatives will usually be considered a threat. Community leaders of a Popperian bent seek to strengthen, defend and deepen the hold of their own systems no less and no less seriously than others. They are monists at rock bottom, but hesitantly so. Their firm belief is that the key to their system's survival, appeal, and vitality is its constant improvement; that one's way of life is best defended and perpetuated not by suppressing, or secluding oneself from other cultures, but by learning from their mistakes, and even more importantly, from their successes. Popperian reflexive skeptics are as serious about their cultures, their normative systems, and their communities as those inclined exclusively towards preservation. If not firmly coupled to constant testing and reassessment, preservation alone, they maintain, is a sure recipe for stagnation, which, in the realm of the social, political, cultural, and even the religious, leads, inevitably in their view, to irrelevance and decay.

On the other hand, keeping an open mind, taking a genuine interest in the ways in which other thoughtful communities conduct their social and spiritual lives, even when radically different from one's own, being willing, in advance, to treat learning *of* others as learning *from* others, should not be confused with taking one's own system lightly. The epistemic hesitance, self-doubt and self-criticism displayed by such communities are not signs of weakness or indifference towards their cultures and traditions. In a deep sense nonchalant unconcern is the enemy of effective criticism and self-criticism. The critical dynamic of Popperian enrichment is rendered all the more effective, the greater the contrast between the participating viewpoints, and the firmer they are initially held. Mutual criticism, and subsequently the learning process that may come in its wake, is more effective the more there is at stake: the more serious the problems can be shown to be, and the deeper the commitment is to the whatever is found wanting, the more pressing the need for a better alternative. The two uncritical extremes render the very process of learning highly unlikely. Dogmatic monists, who hold their views with excessive firmness, will be reluctant to reflect upon them regardless of the problems they may be shown to harbor. And those, equally uncritical relativists, whose views lack real conviction, usually couldn't care less anyway. Ideally, a person or community prudently motivated by Popperian considerations of epistemic humility will seek partners to debate whose views are as seriously held as they are significantly different from their own. And when such considerations are applied to state politics one, it seems, should regard as ideal a multicultural society comprising a plurality of relatively decided communities fairly confident in their views.

But of course all of this can be of no avail unless, to paraphrase Seligman, such a position is indeed articulable from within the assumptions of revealed religion. Epistemic modesty is clearly not a viable option for all religions. If what one believes is truly regarded the God-given truth, the option I am proposing can be of no relevance. Here I can only speak for Judaism, where, I believe, more than the beginnings of such an option are available, if not as readily as one would have liked.

Toward an Halakhically-Committed Pluralism: Principles, Promise and Prospects

It is my firm belief that the formative texts of halakhic Judaism contain clear evidence of a major school of thought whose self-conscious epistemic presuppositions closely resemble those of the Popperian school. The type of approach to which I am referring is aptly described, and, not unlike Popper's own writings, argued in analogy to the epistemic status and growth of scientific knowledge, by one of the most prominent rabbinical figures of the late nineteenth-Century, Rabbi Naftali Tzvi Yehuda Berlin, writing in 1865 or so, in the introduction to his monumental commentary on the Pentateuch: *Ha'amek Davar*.

> Just as it is not possible for the wise student of nature ever to boast knowledge of all of nature's secrets (...), and just as there is no guarantee that what he has accomplished in his investigations will not be invalidated by colleagues (...) who elect to study the same things differently, so it is not possible for the student of the Torah to claim that he has attended to each and every point that claims attention, *and even that which he does explain—there is never proof that he has ascertained the truth of the Torah.* (italics added)

Disappointingly, however, Rabbi Berlin intended his surprisingly Popperian analogy of Torah study to Science to apply only to Biblical exegesis, not to halakha. The model of development he attributes to halakha, offered by him elsewhere,[23] is cumulative rather than revolutionary (resembling

---

[23] See, for example, his commentaries to Exod. 34:1, 27 and Lev. 18:5, his preliminary remarks to Deuteronomy, and most importantly his introductory essay to *She'iltot de-Rav*

mathematical, rather than scientific development). Unlike interpreters of Scripture, the halakhic authorities, he urges, do not seek to invalidate, and improve upon the efforts of his forbears, as does the exegete. In the world of halakha, urges Rabbi Berlin, former rulings are forever binding. Halakhic innovation is achieved not by questioning one's halakhic legacy, but by concentrating exclusively on, and constantly trouble-shooting for, halakhic lacunae.

Rabbi Berlin's two models of development differ in the treatment of their respective legacies. Halakhists, according to Berlin, are wholly bound and obligated by the body of law they inherit, and are, therefore, limited in their legislative efforts to questions on which no ruling had formerly been issued. Biblical exegetes, by contrast, humbled by a profound sense of human fallibility, are encouraged by the tradition itself to challenge and improve upon former interpretations. Owing to their different attitudes to tradition, I have dubbed the two approaches traditionalist and antitraditionalist respectively. Contrary to Rabbi Berlin's depiction—a view widely shared by halakhically committed Jews—I have argued, in a book-length study, for the presence, within the talmudic literature, of a reflective, and insistent antitraditionalist approach *to halakha*, or rather, for the existence of a clear and self-aware tradition of halakhic antitraditionalism.[24] The antitraditionalist voice to which I am referring is detailed and valorized in a variety of talmudic texts and genres. One example will suffice to make my point.

The controversy between the Houses of Hillel and Shammai—the schools of halakhic thought who, according to the talmudic literature, dominated the world of Jewish learning prior to the fall of Jerusalem in 70 AD—is regarded as the paradigm of halakhic disagreement. The talmudic literature contains much legal and legendary material concerning the Houses and the two sages to whom they owe their names. The dispute between the two Houses extends over three-hundred or so matters of halakha. In thirty-four cases, the texts record debates between them following the statement of their initial contrasting views. These allow the reader a glimpse of the different and the shared discursive practices attributed by the texts to the two Houses. One such difference is highly significant. In seven of the fifteen dialogues in which the Shammaites are accorded the last word, the Hillelites are on record as having changed their initial commitment and "admitting" to that

---

Aḥai Gaon, entitled "The Way of Torah", Mossad Harav Kook, Jerusalem, 1986, esp. pp. 5–12. See also Sagi *Elu va-Elu*, pp. 43–6.

24 Fisch, *Rational Rabbis.*

of their adversaries. In none of the eighteen cases in which the Hillelites are accorded the last word do the Shammaites change their mind. A willingness to retract and modify a bona fide halakhic legacy in the face of counter-argument is the hallmark of antitraditionalism. Nothing better attests to such a view than evidence of actual replacement of a former ruling. Unlike traditionalists, who are obligated absolutely by their traditions (and are, therefore, unable to abandon a ruling even when faced with an argument for which they have no answer), antitraditionalists, motivated by perpetual self-doubt, are constantly on the lookout for possible flaws and problems in their own system. It is clear from this and a variety of other evidence,[25] that the talmudic redactors responsible for these texts envisaged the two Houses—whose disputes are formative of the talmudic enterprise—as apt representatives of the two approaches I have been discussing. It is also evident that major talmudic texts that comment on the Houses' controversy, viewed it in the same terms. One such text asserts that the reason we follow the House of Hillel in matters of halakha is because they were humble and accommodating, and would not only give the Shammaite view a hearing, but went out of their way to consider it before stating their own.[26]

The talmudic antitraditionalist is, what I have elsewhere termed a constructive skeptic. He is a critical pluralist who translates his personal intellectual modesty into a powerful method of reflexive critical reasoning. *As a result,* the Hillelites' attitude toward the Shammaites is more than mere toleration, much more. The Shammaites are far more than merely put up with. The Hillelites accord their adversaries more than the right to an opinion, and freedom of speech. No, their interest in *what* the opposition has to say is real. What the Shammaites have to offer *them* is something only a tough-minded adversary can offer: a real and seriously entertained alterantive. They desperately need them around, for the keen challenge they afford their own hesitantly held views. They disagree with them bitterly, but value their presence not for the cozy security of agreement that they offer, but for the exciting, constructive effects of their criticism. They are valued for their profound otherness!

In the light of the problems discussed, talmudic antitraditionalism is clearly an important beginning. The epistemic humility it premises entails two powerful resources for solving the problem at hand: a legitimacy to

---

[25] For further details see H. Shapira & M. Fisch "The Debates between the Houses of Shammai and Hillel: The Meta-Halakhic Issue", (Hebrew) *Iyyunei Mishpat* (*The Tel-Aviv University Law Review*), 22 (3) 1999.

[26] Bavli, *Eruvin* 13b.

rethink halakha as it stands, and, even more importantly for the present context, a healthy, religiously meaningful critical pluralism that allows one to both keenly disagree with, and, at the same time, seriously value the existence of halakhic opinions opposed to one's own. As it stands at present, halakha, I have argued, forbids actively defending and assisting halakhically objectionable forms of conduct. Reclaiming the voice of halakhic antitraditionalism is to reclaim the license not merely to rethink former rulings, but to rethink them within the desired environment of a vigorous and authentic meta-halakhic pluralism—the very attitude we found so lacking in the current mindset of halakhic Judaism.

But let's not get carried away. Interesting as all this may be for the study of talmudic discourse and culture, the immediate resources it offers for the type of latter-day concern discussed in the present volume, remain, it would seem, very limited. For one thing, anything to do with meta-halakhic deliberation remains, by definition, an in-house affair. Tolerating diversity within the community is always a matter very different from the way the community as a whole faces the world without. Furthermore, the 'radius of toleration' within which the Hillelites' latitudinarianism in fact asserts itself in relation to the Shammaites is disappointingly narrow in comparison to the degree of ethical and cultural diversity latter-day halakhically committed Jews are challenged by. The immediate reaction of almost all the students of the talmudic literature to whom I have presented this argument has been little more than a polite dismissal.[27] Halakhic pluralism, they assert, is limited to, and wholly contained within the boundaries of

27 Sagi's book-length study of post-Talmudic attitudes to halakhic disagreement shows awareness to the Popperian nature of the the position I term antitraditionalist, though almost as an afterthought He distinguishes between monistic and pluralistic approaches to halakhic disagreement (the former treating their own positions as true and their adversaries as mistaken, the latter denying there being one solution to a given problem (pp. 88–9)) and goes on to classify the entire literature in terms of variants of the two positions—none of which premise the type of self-critical, modest monism I have been advocating. Then, in the last chapter but one, he does, introducing a 'weak pluralism' he associates with Popper, Mill and Voltaire, to which, he claims two authors previously classified differently, can be shown to ascribe (pp. 194–5). However, even here the entire discussion remains confined to halakhic disagreement, and is not extended to challenges from outside the fold. On this he is explicit elsewhere: because of its inherent self criticism the price a religious weak pluralist has to pay is high, for he runs the constant risk of his religious beliefs being refuted. "Furthermore", adds Sagi, "weak pluralism" does not sit comfortably with the language and practice of halakha" ("Jewish Religion", pp. 189–90). He is obviously right on both claims regarding the attitudes of latter-day halakhically committed Jews. With respect to the talmudic literature, however, I strongly disagree. For similarly disappointing conclusions also see A. Ravitzky, "The Question of Tolerance in the Jewish Tradition: Between Pluralism and Paternalism" (Hebrew), A. Sagi and Z. Safrai

acceptable halakhic opinion. "These as these are the words of the living God", states the Talmud in summarizing the three-year controversy between the Houses, I am constantly reminded. To extrapolate from the contained and muted pluralism practiced by the Hillelites toward other God-fearing, equally observant communities of Jews, to modern multiculturalism, I am told, is not only wrong-headed but irresponsible.

Had I intended to argue apologetically that by listening carefully to the talmudic texts of late antiquity one could find a ready-made solution to a decisively latter-day problem, I would have well deserved all of that and more. What I am pointing to in this literature is not a prophetic premonition of current Western multiculturalism, but an epistemic attitude; not something the rabbis couldn't have been aware of (and had they been would have probably balked), but something of which they seem to have been perfectly aware, and totally at ease—namely, a profound sense of the limits of humankind's capacity to know, resulting in a genuine and proven epistemic modesty which they applied *even* in the realm of firmly held halakhic legacies! This is not to be taken lightly, especially if one recalls that, having at their disposal no procedure for deciding between conflicting halakhic legacies, traditionalists are obliged to either refrain from deciding altogether, and to leave the matter inevitably open (until the "return of Elijah"), or reluctantly to allow each member of the community to abide by his or her own tradition—in which case the school will irreversibly be split in two. "Thus," states the very traditionalist Tosefta, *Sanhedrin* 7:1 "as the disciples of Shammai and Hillel who had not attended upon their Masters sufficiently increased in number, so increased the number of halakhic disputes in Israel," and, adds the version cited in the Bavli, "the Torah is rendered (split into) two (separate) *Torot*."[28] What I urge we listen to and extrapolate from is the very principle of epistemic modesty operating at the heart of talmudic antitraditionalism, and the manner in which it provides the hard, normative core of halakhic Judaism—the halakhic system itself—with a powerful, decidedly open-ended self-correcting mechanism.

(eds.), *Between Authority and Autonomy in the Jewish Tradition*, Hakibbutz Hameuchad, Tel-Aviv, 1997, pp. 396–420.

28 Tosefta, *Sanhedrin* 7:1 and *Hagiga* 2:4, cited Bavli, *Sanhedrin* 88b. See also Bavli, *Sotta* 47b and Yerushalmi, *Sanhedrin* i, 19a. Yerushalmi, *Hagiga* ii, 77d brings the point home the most forcefully of them all: "Thus, as the disciples of Shammai and Hillel who had not attended upon their Masters sufficiently increased in numbers, and the number of (halakhic) disputes in Israel increased, they became two (separate) sects—these would rule a thing unclean while these would rule it clean, and it will not be settled until the coming of The Son of David (i.e. the Messiah)!"

It is not a solution to the problem at hand. Much conceptual and halakhic work has yet to be done before a solution can be boasted. But in the anti-traditionalist voice of talmudic Judaism the necessary conceptual and philosophical framework exists for the potential development of an original, halakhically-commited, Jewish version of the fourth, Popperian option for which this paper has been arguing. This is the important resource that the formative texts of rabbinic Judaism have to offer their latter-day heirs who, for the first time in history, face the great challenge of modern statehood.

JUDAISM AND THE RELIGIOUS CRISIS OF MODERN SCIENCE[1]

Menachem Fisch

Introduction

From the advent of the scientific revolution in the mid-sixteenth century onwards, the ethos and mindset of modernity and of modern science have evolved hand in hand, bringing in their wake a mounting secularism, and with it the now familiar tensions between science and religion. By the end of the nineteenth century, the strained relations between science and religion had acquired the dimensions of a full-blown crisis. Many tend to ascribe the crisis to the marked discrepancies between what scientists were willing to say about the world and what devout Christians believed their religion required them to believe. Discrepancies of this sort certainly did emerge, some quite dramatic, but it is extremely hard to imagine that they alone could be held responsible for the full extent of the crisis. What developed was far more than mistrust, or even condemnation expressed by devout Christians toward developments in the life sciences and cosmology. In fact, the late enlightenment witnessed far less religious suspicion of science, than a mounting, scientifically motivated exasperation with religion and religiosity. And it is this latter tendency that strongly suggests that much more was, and remains at stake than a mere conflict of truth claims. Modernity's scornful dismissal of religion and religiosity, along with the reciprocal reaction to which it gave rise, runs much deeper than any prosaic disagreement as to, say, the validity of geological dating can reasonably account for.

In the first part of this chapter, I offer, in very brief outline, a different kind of explanation for the emergence of the modern tensions between

---

* This chapter first appeared in: Nature and Scripture in the Abrahamic Religions: 1700–Present, in Brill's Series in Church History and Religious Culture 37 (2008). Koninklijke Brill NV.

[1] Various versions of this paper were presented as the Selig Brodetzky Annual Lecture, Leeds University (May 2002); at a conference on scientific and religious interpretation, Haverford College (June 2002); as Templeton Lectures at Bar-Ilan University (March 2003) and Utah State University (October 2004). An early version was published in Hebrew in Halbertal and Sagi 2005.

science and religion—one that stresses its philosophical and theological dimensions, centering on the conflicting conceptions of human rationality that the work of prudent 'interpreters of nature and Scripture' were seen to premise, rather than on the conflicting pictures of the world that they came to paint. I shall be interested, therefore, less in what scientists were saying about the world during that period, than in the philosophical attempts to understand the nature of the enterprise and achievement of science; less in the history of science, that is, than in the history of the philosophy of science. And it is against this backdrop that in the second, and main part of the paper, I shall turn to consider how Judaism can be said to fare in this respect.

The Newtonian Moment

A crucial turning point in the history of the relations between early modern science and religion was the publication in 1687 of Sir Isaac Newton's great work of physics: *Philosophiæ Naturalis Principia Mathematica*. The *Principia* was by all counts an astonishing scientific achievement, but its importance for the story I wish to tell lay less in its scientific content than in the acute challenge that it posed to philosophy, for Newton's *Principia* changed the very nature of philosophizing about science. Prior to Newton, philosophy was primarily concerned with the project of establishing science properly. Francis Bacon's *Novum Organum* (1620) and René Descartes's *Discourse on Method* (1637) and *Meditations* (1641) stand out in this respect. Each presented a radical, and meticulously argued (yet very different) appeal to begin the study of nature from scratch, and to establish it anew upon sound philosophical foundations. But with the publication of Newton's physics, the project of establishing science was rendered obsolete. From that point on philosophy of science was rendered an essentially interpretative rather than a constitutive undertaking. If for Bacon and Descartes the task of philosophy of science was to legislate epistemologically and methodologically for a reliable (and as I shall argue, religiously viable) interpretation of nature, after Newton, famously for Kant, the question for philosophy became that of making philosophical sense of the particular interpretation of nature that Newton had fashioned. The interpretative turn in philosophy of science reached completion in William Whewell's monumental *Philosophy of the Inductive Sciences* (1840) significantly subtitled *Founded on their History*, referring to his three-tome *History of the Inductive Sciences from the Earliest to the Present Time* published three years earlier. My thesis will be that this major shift of philosophical interest—from the problem of explaining how one ought to read the book of nature, to that of justifying how it was in fact being read—marks more than a major turning point in the history of philosophy. It also marks the point at which the formerly harmonious relationship between science and religion took its significant first turn to the worse. But to appreciate it as such we must briefly return to the great upheaval in European Christendom initiated by Martin Luther almost a century and half before Newton.

The Bible, Protestantism, and the Rise of Natural Science,[2] *and Beyond*

The Protestant Reformation rendered the Holy Scriptures an open book. 'The Meaning is in the Text' was the rallying cry of the early Protestant reformers, who urged their followers to discover it for themselves. Direct access to Scripture, and with it, unmediated access to the very Word of God was granted to all willing to approach the Bible with an open mind and an open heart. Early Protestantism, one could say, privatized divine revelation, transforming it into a personal matter of prudent reading, the very prudence of which guaranteed success.[3] Scripture does not conceal its message, but speaks plainly and directly rendering the divine word accessible to all who seek it truly and purely. Therefore, reading Scripture does not require the mediating services of church doctrine, clerical go-betweens or priestly exegetes. But if all believers were to be able to discern Scripture's meaning and message for themselves, it had to be assumed that they resided not merely 'in the text,' but in the immediate most literal sense of its words.

In *The Bible, Protestantism, and the Rise of Natural Science*, Peter Harrison makes the further, and novel claim that the Protestant reformers' insistence on literal interpretation played a major role in the emergence of early modern science, by virtue of the new meaning that was lent to the old and religiously appealing two books metaphor. In the hands of the early reformers, argues Harrison, the two books metaphor was extended beyond the simple idea of a common authorship, and away from the more subtle idea of one book holding the key to the other.[4] For the early reformers the two

---

2 Harrison, 1998.

3 As Harrison puts it: "In much the same way that the faithful were granted direct access to Scripture, they were granted unmediated access to God himself" (1998, 99).

4 "For Bacon," notes Moore, "the book of God's works is 'a key' to the book of God's words; students of nature may therefore instruct interpreters of the Bible. To reverse the

books metaphor was rendered a powerful and suggestive analogy between the way the two texts were conceived. As Kenneth Howell nicely puts it, Harrison reversed the common view that the new sciences provoked new interpretations of the Bible, maintaining that sixteenth-century people "found themselves forced to jettison traditional conceptions of the world" when they "began to read the Bible in a different way."[5]

Reading Scripture became the model, rather than the key, for studying nature, bestowing upon science much of the new urgency, enthusiasm, and promise it was enjoying. This explains the impact of the Reformation on the rise of natural science. But Harrison's most novel and important claim is that the early reformers' attitude to the Bible not only motivated the early modern turn toward nature, but inspired an importantly new conception of nature. By insisting that the Bible's message resided in the plain, literal meaning of the text, the reformers were forced to empty the biblical text of all manner of symbolical, metaphorical and allegorical meaning. And by analogy, so was nature. By virtue of its new understanding of two books metaphor, argues Harrison, Protestantism was responsible for rendering nature, for the first time, purely *factual*.

Harrison shows as never before how, through the mediation of the two books metaphor, early Protestantism revolutionized natural science by revolutionizing Bible study; how interest in nature was lent religious significance, and at the same time, and for the same reason, how it shifted focus away from what natural phenomena symbolize, even religiously, to the phenomena themselves—to their causes, and to the complex relations that obtain between them. But Harrison, I believe, overlooks an important aspect of the story he is the first to tell. By concentrating exclusively on the texts themselves, as it were, on how for the early reformers the book of God's word and that of his works came to resemble each other, he fails to give due notice to the assumptions they made, and conclusions they drew regarding the act of reading them. Or to use a more philosophical turn of phrase, Harrison concentrates more on the ontological, than on the epistemological and methodological consequences of his novel thesis.

In the personal invitation to discover the Word of God for themselves, extended in Luther's term to 'the priesthood of all believers,' lie the religious roots of modern individualism and modern selfhood. And in the

---

relationship and attempt to make things in nature conform to words in the Scriptures would be to 'unwisely mingle or confound these learnings together'." (Moore 1986, 323 quoting Bacon 2000, Bk.I, VI.16 and I.3).

5 Howell 2002, 7, quoting Harrison 1998, 4.

Protestant reform of biblical hermeneutics, lie the roots of a particular modern notion of inquisitive prudence dedicated to unbiased, detached, meticulous inquiry; one that combines diligence and an open mind—which really means a largely emptied and passively receptive mind; a mind open to listening attentively to what the text has to say, while abstaining as far as possible from making its own interpretive assumptions; a mind dedicated to freeing itself of prejudice and giving itself piously to reading *out* the true message of Scripture, while firmly resisting the temptation to read one *in*.

If when carried over from Scripture to science, "the assertion of the primacy of literal reading" entailed "a new, non-symbolic conception of the nature of things,"[6] then, by the same token, the insistent "literalist mentality" demanded of the readers of Scripture, entailed an equally insistent *empiricist* mentality with regard to the study of nature. If, like religious truth, scientific truth was also believed to reside in nature, ripe and ready for human discovery, in the immediate, 'literal' sense of what meets the human eye, then similar to Scripture, nature too could and should be made to reveal her secrets by prudently reading them out inductively, unmediated by hypothesis or theory. And, therefore, just like the study of Scripture, neither should science require the mediating services of clerical go-betweens or priestly exegetes—especially not those of "the Aristotelian Church" to use Luther's deriding formulation.

Bacon and Descartes's Religiously Motivated Philosophies of Science

One could, therefore, speak of the emergence of an early modern Protestant *image*, or *philosophy* of science that endorsed an analogous notion of epistemic prudence that valued clever, diligent and unimaginative minds seeking passively to receive nature's plain 'literal' message, while carefully abstaining from speculating as to what it might be. Here, I submit, extending Harrison's thesis by going a philosophical step beyond him, lie the Protestant roots of early modern empiricism and the experimental method it inspired—a philosophy of science that found exemplary expression in the work of Francis Bacon.[7] The sobering analysis of the methodological

[6] Harrison 1998, 114.

[7] On Bacon's Protestantism see Jardine and Stewart 1998, esp. chap. 2–4. On his philosophy of science see Urbach 1987. James Bono fails to make the connection between the factualism underlying Bacon's account of the 'book of nature' and the early reformers' literalist biblical hermeneutics, however, his analysis of Bacon's approach dovetails

empiricist[8] point of departure (the four idols), and the highly original eliminative inductive methodology of Bacon's *Novum Organon*,[9] would be adopted thirty-five years later by Robert Boyle and his eleven colleagues as their 'official' philosophy when they established the Royal Society in November 1660. Thomas Sprat, celebrating the new, and very Baconian experimental method in his *History of the Royal Society* of 1667, waxed poetical in the most Protestant terms of there being two Reformations, each prizing the original copies of God's two books while bypassing the corrupting influence of scholars and priests![10]

---

perfectly with the extension of Harrison's thesis proposed here. If, as Bacon maintains, God's "Book of Nature [is] written not in symbols, but in things themselves," he asks,

> What does it mean for man to "read" that text? ... For one thing, it means that the "meaning" of that text is inherent, or immanent, to itself. The fundamental units that make up the text of nature do not acquire their intelligibility through metaphoric relations to other units and, ultimately, to other texts.... The language of things in which God has composed His Book of Nature instead embodies and enforces a metonymic order of meaning. Such an order grounds the intelligibility of nature in the contiguity and spatiotemporal relations between and among things. (Bono 1995, 233–4)

[8] I qualify Bacon's empiricism as 'methodological' to indicate that unlike later British empiricists (notably Locke and Hume), Bacon's was a philosophy of scientific knowledge, not of knowledge in general. He nowhere addresses the question of the ultimate sources of knowledge, and offers no general theory of the formation of ideas or concepts. Like Luther's literalism, Bacon's empiricism is prescriptive—limited to what *should* be allowed to bear on science's picture of reality.

[9] For an impressive and systematic latter-day adaptation of Bacon's eliminative inductive method see Cohen 1970; 1977, and 1989, esp. ch. 5.

[10] I quote the paraphrase of Sprat in Brooke 1991, 111. Sprat's original text (cited by Harrison 1998 as one of the mottos of Chapter 3 entitled "The Two Reformations") reads as follows:

> ... the *Church of England* will not only be safe amidst the consequences of a *Rational Age*, but amidst all the improvements of *Knowledge*, and the subversion of old Opinions about *Nature*, and introduction of new ways of Reasoning thereon. This will be evident when we behold the agreement between the present *Design* of the *Royal Society*, and that of our *Church* at its beginning. They may both lay equal claim to the word *Reformation*, the one having compass'd it in *Religion*, the other purposing it in *Philosophy*. They both have taken a like cours[e] to bring this about; each of them passing by the *corrupt Copies*, and referring themselves to the *perfect Originals* for their instruction; the one to the *Scripture*, the other to the large volume of the *Creatures*. (Sprat 1667, 370–1)

The extension of Harrison's thesis proposed here as to the emergence, the basic fault lines and development of early modern philosophy of science in its relation to religion may also be seen as an extension of Simon Schaffer and Steven Shapin's account of the Boyle-Hobbes controversy (Shapin and Schaffer 1985). They argue that the dispute between Hobbes and Boyle was as much a dispute about politics as it was about the latter's admittedly Baconian philosophy of science. The extension of Harrison's thesis proposed here purports to add a third, religious dimension to Shapin and Schaffer's two.

But in so extending Harrison's Protestantism and science thesis to epistemology and methodology, its scope is inevitably narrowed from an account of the religious roots and context of early modern natural science in general, to a thesis about the religious roots and context of a particular early modern philosophical approach to the study of nature—namely, early modern (methodological) empiricism. Harrison writes as if the literalist mentality of the Protestant reformers could be held responsible for the equally impressive surge of such rationalist approaches to science as Galileo's[11] and especially Descartes's.[12] This, I believe, is the one weakness of his otherwise superb book.[13]

Alongside the rise of Bacon's empiricist philosophy of science and its acceptance by Boyle and his colleagues, early modern Europe witnessed an equally confident, equally enthusiastic and equally urgent development of a systematic scientific rationalism—a philosophy of science that lay special emphasis on mathematical physics, and found exemplary and systematic expression in the philosophy of René Descartes. Descartes's rationalist philosophy of science owed its origin and inspiration, not to Protestant attitudes to the interpretation of Scripture, but, as Stephen Toulmin, Stephen Menn, and others suggest,[14] to the Catholic reaction it provoked, and, to extend Harrison's argument, to the analogous function of the two books metaphor in a Catholic setting, where it was inevitably attributed a very different meaning.

---

11 Harrison 1998, 267–8.

12 Ibid., 263–4.

13 To put the point somewhat more schematically, for Harrison, the two books analogy serves to ground an argument from how the early reformers envisioned the proper approach to Scripture, to how nature came in fact to be studied during that period. This cannot easily be argued for all of science (or in the language of the day, all of natural philosophy)—especially not for mathematical physics and astronomy. Unlike Harrison, the proposal of this paper is to read the two books analogy as one between two normative and constitutive claims, one prescribing the desirable way to approach Scripture, the other prescribing philosophically the desirable way to do science. Interestingly, at crucial junctures of his argument, sensing the difficulty perhaps, Harrison switches from using the term 'natural philosophy' to using 'natural history.' Cf. Harrison 1998, 95–7.

14 While more interested in contrasting the "unrelenting" rationalism of Descartes with the "modest" humanist reasonableness of Montaigne than with that of Bacon (198ff.), and although he considers Bacon's philosophy (along with Montaigne's) as a product of renaissance humanism rather than of the Reformation (36–7), Toulmin 1991 sees in the Counter Reformation's "hardening of Catholic teachings" the immediate context relevant to Cartesian rationalism (69–80). For similar assessments of Descartes, see Menn 1998, 53ff., and Bono 1995 esp. ch. 8.

Counter Reformation Catholicism would have nothing of Lutheran literalism. For Catholics the biblical text was considered replete with deep theological significance that required expert interpretation and decoding. The idea that the word of God could reside in the simple meaning of the text, and be read off it directly by anyone was unthinkable. Catholic interpretation of Scripture required the mediation of a firmly established metaphysical system of rational theology that demanded long and laborious training, and unflinching prior commitment. This had always been the case, but, as Toulmin explains, once locked in confrontation with the protestant reformers, especially after the Council of Trent, "Catholic teachings hardened into theses (or 'dogmas') that were no longer open to critical discussion … and whose immutable truth it was politically indispensable to assert … Instead of free-wheeling *Summas*, the seventeenth century was fed a diet of centrally authorized *Manuals*;"[15] Lest no mistake be made, the Fourth Session of the Council of Trent decreed explicitly that

> in order to restrain petulant spirits … no one, relying on his own skill, shall—in matters of faith, and of morals pertaining to the edification of Christian doctrine—wresting the sacred Scripture to his own senses, presume to interpret the said sacred Scripture contrary to that sense which holy mother Church—whose it is to judge of the true sense and interpretation of the holy Scriptures—hath held and doth hold.[16]

In its Catholic version the two books analogy would hence reason that just as interpreting Scripture requires prior assent to an elaborate system of necessarily true rational theology, so the study of nature demands the prior elaboration of and assent to an analogous system of absolutely certain scientific truth—which for Descartes were the truths of mathematics.[17] According to Stephen Menn's interesting reading of the return to Augustine evident in Descartes's mature philosophy, one finds in the *Meditations* far more than a Catholic counterpart to Bacon's philosophy of *science*. The *Meditations*, he argues, "is addressed not to philosophers but to theologians,"

---

15 Toulmin 1991, 77–8.

16 *The Canons and Decrees of the Sacred and Oecumenical Council of Trent*—Session the Fourth, 19.

17 Stephen Menn nicely compares and contrasts Descartes's "*mathesis, ou mathématique universelle*" and Bacon's "regime of facts" as radically new (if diametrically opposed) "*de-in-scriptive* hermeneutics of nature" that stand in conscious opposition to "the textual, symbolic, "emblematic, and narrative strategies" formerly carried over from the "hermeneutics of exegesis" by those who read the two books as one. What Menn fails to notice, however, is the extent to which, despite their radical novelty, the two new philosophies of science remained centrally faithful to their respective religious settings.

and is offered to them "as *the true philosophy of the Catholic Reformation*" in general.[18]

> The *Meditations* argue first that we cannot know bodies unless we first know God and the soul (Meditation 1), then that we can know God and the soul independently of any knowledge of bodies (Meditations 2 and 3), and finally that we can derive a knowledge of bodies from this pure knowledge of God and the soul (Meditations 4–6). Augustine's metaphysics of God and the soul is a possible candidate for Descartes' foundational discipline because it does not depend on any doctrine of the physical world, and indeed results from a discipline of *aversio* from the senses. So Descartes can base his physics on his metaphysics without danger of circularity.[19]

In other words, argues Menn, Descartes's rationalist "foundational discipline" is presented as prior to, and constitutive of the proper Catholic reading of *both* books.

Bacon's Reformation, Protestant account of science, if I may call it so, insisted on grounding scientific truth in the immediacy of factual experience from which one then reasons *up* inductively to theory. Descartes's Counter Reformation, Catholic science, by contrast, insisted, equally vigorously, that it be grounded in the *a priori* certitude of the mathematicized systems of ideas given by pure reason, from which one then reasons *down* deductively to the facts. The former envisaging the discovery of scientific truth as an elaborate, if hesitant, 'literalist' process of reading it *out* of the empirical data, the latter, by contrast, as an equally elaborate process of 'doctrinal' reading *in*.

The Parting of Ways

If there is any truth in my proposed extension of Harrison's thesis, then the dramatic rise of science during the seventeenth century owed its inspiration and philosophical rationale not to one, but to two powerful foundational philosophies of science that owed their origins to radically different versions of the two books analogy, that were grounded respectively in radically different Reformation and Counter Reformation approaches to Biblical interpretation. Most importantly, in the framework of both philosophies, science and religion meshed naturally and reinforced each other fruitfully and harmoniously. Both Baconian inductivism and Cartesian

18 Menn 1998, 53.
19 Ibid., 209.

rationalism promised lofty, ambitious and religiously meaningful visions of grand scientific accomplishment, backed by foundational and systematic philosophical reasoning. And needless to say, they contradicted each other completely.

Viewed thus, the tensions that later developed between science and religion cannot be attributed to incongruities between inherently contradictory undertakings whose mutual exclusion was predetermined. The original appeal and harmony between the constitutive philosophies of science fashioned by Bacon and Descartes and their respective religious cultures was deep, meaningful and authentic. One cannot easily argue that the crisis that developed thereafter proved in retrospect that the initial harmony was an illusion. It was not. To explain how and why the crisis developed is to explain how and why the original twofold harmony began to wax and wane; how it became gradually undone and eventually turned sour. It is a complicated story that in the present context can be outlined only very briefly.

The sweet harmony between religion and science attested to by both early modern philosophies of science proved extremely short-lived, not because it was disproved, but because of a profound scientific development. As I mentioned at the outset, the crucial turning point in this respect was the publication of Newton's great work of physics in 1687. In a word, Newton's *Principia*, justly upheld as a supreme scientific achievement, proved to be a philosophical nightmare, for it exhibited a perplexing mixture of mutually irreducible empiricist and rationalist elements that neither philosophy of science was capable of accommodating from its own purist perspective. Normally, when confident subscribers to a well established philosophy of science find a work of science to rudely defy their prescriptions, they will tend to declare it confused or incoherent.[20] But in the case of Newton's physics such a response was out of the question. The solid and unanimous reception of Newton's physics as an unprecedented scientific accomplishment rendered it virtually immune to philosophical criticism and ruled out any possibility of such philosophic reproach. Both philosophies of science were henceforth forced to make substantial concessions to the new physics that swiftly came to lay significant strain on the firm bonds they bore to their respective religious contexts. The full disengagement of late enlightenment understandings of science from religion would come about in two

20 And explained away as the sorry, irrational or a rational result of 'external' circumstance. See Lakatos 1978.

main stages, but by the end of day nothing would be retained of the two books analogy that prompted the scientific visions of Bacon and Descartes, and nothing of the fruitful and reinforcing enmeshment of religion and science out of which the philosophical understanding of the scientific revolution initially set forth.

Despite the *Principia*'s axiomaticized Cartesian form and heavily mathematicized presentation, it was Cartesian rationalism that would be the first of the two positions to give.[21] In the course of the eighteenth century, Descartes's original rationalism virtually vanished from sight, giving way to two very different accounts of the mind's constitutive contribution to scientific knowledge: the so-called instrumentalism of Lagrange's analytical mechanics (rendered a century or so later a full-blooded philosophy of science by Pierre Duhem),[22] and the constructivism of Immanuel Kant[23]—neither of which could be considered a viable scientific correlate to a Catholic conception of biblical interpretation. Instrumentalism denied physical theories any claim to truth, viewing them merely as 'instruments' construed and perfected to economically organize and store, and effectively predict empirical phenomena, with no pretension to depict reality. Kant, by contrast, granted science its traditional realist and explanatory functions, as well as its claim to truth, but denied it any access or insight into external reality "in itself." With instrumentalism renouncing truth as the goal of science, and Kant, external reality as its subject-matter, both positions presented radically sobering retreats from Descartes's heady rationalist vision of scientific achievement.

Bacon's 'literalist' inductivism proved more resilient, and was still commanding the assent of such prominent early nineteenth-century figures as John Herschel and John Stuart Mill. But their contribution to the understanding of science as it was practiced at the time would be at best marginal. Mill and Herschel's inductivism will be remembered in the history

---

21 Apart from its incapacity to accommodate the more inductivist and empirical portions of Newton's physics (in particular Book III of the *Principia*, and the "proofs by experiment" systematically employed throughout the *Opticks*, let alone the notorious "Rules of Reasoning in Philosophy [i.e., Science]" situated between the *Principia's* Books II and III), Descartes' rationalism faced insurmountable difficulties accounting for the differential and integral calculus fashioned by Newton and Leibniz that mathematically grounded Newtonian physics. As Berkeley (1734) amply proved, the calculus fell considerably short of exhibiting the self-evident certitude (clarity and distinctness) required by Descartes's philosophy.

22 See especially Duhem, 1954. For explicit praise of Lagrange's 'instrumentalism,' see Duhem 1980 esp. ch. 8, and Dugas 1988, 456f.

23 For a more detailed account of these developments see Fisch 1989.

of philosophy more as pledges of allegiance to the philosophical ways of old than as honest and serious attempts to make philosophical sense of the highly developed physical sciences of their day. By contrast, the new empiricists—beginning with David Hume, continuing in the work of William Whewell and finally Karl Popper advocated philosophies of science that retained nothing of Bacon's 'Protestant' literalism of old.

In the heady wake of Newton's great achievement, philosophers found themselves having to abandon major elements of the two philosophies of science that had launched modern science on its way. As these concessions were made the intimate ties that, by virtue of the two books analogy, had briefly bound the budding culture of enlightenment science to those of Reformation and Counter Reformation Christianity respectively, were gradually undermined.

The crisis developed in two main stages. First to go was the two books analogy itself, which to the best of my knowledge received no mention (in the Harrisonian sense in which I have been using it) after 1700 or so, except perhaps in the rhetoric of some of the British natural theologians. The study of Scripture and that of nature simply ceased to be thought of as analogous undertakings. Science and religion were each still identified in terms of the different 'readings' to which they lay claim and held true, but in the light of the standard set by Newton, neither the kind of knowledge scientists sought to produce, nor the process of inquiry by which they succeeded in doing so could be described philosophically in a manner analogous to that in which Protestants or Catholics were expected to read Scripture. Scientific enquiry came to be conceived of (differently by the different schools of philosophy) as a unique form of knowledge-acquisition, that bore little resemblance to either of its religious counterparts. As a form of humanly possible intellectual achievement, science came to be considered a class of its own.

The second stage proved to be the crucial one, for it concerned rationality itself. Again inspired by religion, early Enlightenment visions of human rationality measured it in terms of epistemic trustworthiness. To act rationally was to act reliably, on the basis of trustworthy knowledge. Science was considered a paradigm of rational endeavor precisely because it was thought to amass great bodies of truth by applying valid, and truth-preserving methods of reasoning and computation to reliable and trustworthy data. The same could be said of religious truth. The methods by which science acquired truth were by now considered different from those of religion, but difference of itself does not necessarily breed contempt. But by the second half of the eighteenth century even this image of scientific knowledge was becoming undone, and with it, in the nineteenth century came the final "collapse of the complex system of similitudes which had characterized pre-modern knowledge."[24] It was a complex process to which several factors contributed, among them, the development of formalist approaches in mathematics and instrumentalist approaches in mathematical physics mentioned above (both initiated by Lagrange, the former in his *Théorie des fonctions analytiques,* of 1797, the latter in his celebrated *Mécanique Analytique* 1788); the emergence of the historical sciences of emergence in astronomy, geology, the life sciences and the sciences of man, whose theories were incapable, in principle, of ever boasting the same levels of credence and reliability as the "actualist" theories of the physical sciences; and the centrality of counter-factual reasoning in science (e.g., in justifying the law of inertia) for which the very notion of empirical reliability seems inapplicable. These new scientific tendencies, developments, and realizations, brought to the fore the inherently uncertain, speculative and critical dimensions of scientific research for which no parallel existed in contemporary Christian self-images of religious knowledge. In the concluding paragraph of his book, Harrison puts the point somewhat differently. In post-Reformation Christianity

> The mainspring of religion ceased to be the performance of ritual acts... Instead, religion came to be identified with systems of belief. Various 'religions' were thus distinguished by the beliefs of their adherents. The seventeenth century witnessed a dramatic reversal. 'Science,' prior to this, had largely been a matter of subscribing to particular doctrines; Christian religion, on the other hand, had been to do with the performance of certain activities. Now science [had] become primarily an activity, while religion... a matter of subscribing to unchanging dogmas.[25]

Science had not only become an activity, but one whose self-image was changing rapidly, as it became more and more apparent that the two early enlightenment 'trustworthiness' notions of scientific rationality were a conceit. The acquisition of scientific knowledge, it was realized, involves much more than applying truth preserving rules of inference and computation to reliable data. The creation of scientific theories, argued philosophers like William Whewell and later Karl Popper, requires imaginative speculation and bold hypotheses, that can be rigorously tested though never proved true.[26] Science continued to be hailed a paradigm of rationality, and justly

---

[24] Harrison 1998, 273.

[25] Ibid.

[26] Whewell believed that although no amount of scientific testing could prove a theory true, some scientific theories are nonetheless capable of proving of themselves

so, but the very notion of rational endeavor had changed. The rationality of science, philosophers had come to realize, is not a matter of proven confidence in what science believes, so much as a humble, self-doubting, yet constructive awareness of its liability to err.[27] Science advances by relentless trial and error, rather than by accumulating well-validated findings, and is rational by virtue of the robust and constant self-criticism by which it proceeds, rather than by its alleged ability to vouch for the truth of its findings. As James Frazer put it concisely in the 1890 first edition of *The Golden Bough*: "After all, what we call truth is only the hypothesis which is found to work best."[28]

To put Harrison's point slightly differently, Bacon and Descartes's philosophies of science grounded the truth of scientific theory in an appeal to an 'external' source of foundational true 'data'—the former in the carefully accumulated and meticulously analyzed three tables of empirical data, the latter, in the divinely vouched for mathematical truths of pure reason. Both philosophies, one could say, were revelatory in this respect, in a manner deeply analogous to the interpretation of Scripture in their respective religious settings. From the late eighteenth century onwards philosophers of science (with the uninteresting exception of such British empiricists as Herschel and Mill) ceased to view scientific knowledge in these terms. Theories were described as highly imaginative conjectures that are tested against data deemed reliable, but not as grounded in or constructed on the basis of such data. It is at this point that the interpretation of nature ceased to be conceived of as a revelatory undertaking in the sense described, and the very image of the two books collapsed even as a metaphor.[29] Protestants and Catholics continued to lay claim to, and be identified with vast bodies of religious truth by submitting themselves to the 'external' sources of

---

true in retrospect. True theories distinguish themselves from merely good ones first, in repeatedly exhibiting what he termed 'consilience of inductions,' or explanatory surprise, and second, in generating 'progressive simplification' as they are repeatedly applied to new and different kinds of phenomena. In Whewell's opinion both gravitation theory and the undulatory theory of light had proved themselves true in this manner. Both criteria are as questionable (as criteria of truth according to Whewell's own philosophy!) as they are intriguing (as criteria of acceptability), and have attracted considerable attention. For further details see Fisch 1991, 147–54.

27 For a comprehensive study of this development during the nineteenth century and its relation to religion, see Herbert 2001.

28 Frazer 1890, 212, cited by Herbert 2001, 2.

29 For a recent account of secularization in terms of relocating "the place of fullness," from an external "power that goes beyond us" to the immanence of "self-sufficient reason," see Taylor 2007, 4–22.

religious authority. Science, by contrast, gradually ceased to be identified with the systems of knowledge it erected, and came to be thought of, primarily, as an activity, as a rational quest, in the self-critical, self-sufficient sense of the term, in the course of which, quite unlike religion, its forever tentative claims to truth are persistently questioned and modified. The great divide between science and religion in the Christian world was thus rendered complete: if science is what human rationality can amount to, then religion is not merely different, but downright irrational! On such a showing, modernity's great challenge to religion has, therefore, far less to do with questions concerning the age of the universe, or with squaring neo-Darwinism with Scripture, than with the accusation that in submitting to their faith's demand for epistemic commitment, religious individuals were surrendering something central and constitutive of their very humanity: the healthy robust self-questioning of the truly rational, that finds paradigmatic expression and warrant in the shining example of modern science.[30]

How does the Jewish tradition fare in this respect? In the remainder of this essay I shall make the case for the existence within Judaism's formative canon of late antiquity of a clear and self-conscious voice whose attitude to religious learning, commitment, obligation, and even performance, remains intriguingly congruent with the epistemic virtues and view of rationality associated with modern science's self-image, despite the fact that traditional Judaism attributed no religious value to the study of nature per se. My argument will not be that the view of religiosity I shall be exposing ever manifested itself fully as a lived religious reality (although the clear expression it finds in the texts I shall be considering surely attests to the existence of real individuals whose view it was). What follows is a chapter in the history of Jewish ideas of religiosity, with a view to exposing a voice that from the latter-day perspective of religious studies seems no longer to be counted as a viable conception of religiosity at least among the Abrahamic religions. But I am getting ahead of myself.

---

30 Nietzsche's views on religion are well known, his view of science, less perhaps. According to Richard Rorty, René Berthelot was perhaps the first to note the interesting affinity between Nietzsche's perspectivism and William James's pragmatist theory of truth. See Nietzsche 2001, §354 where he notes: "We do not even have any organ at all for *knowing*, for truth; we 'know' … just as much as may be *useful* in the interest of the human herd." Quoted by Rorty 2007, 27.

Interpreting Scripture and Nature—A Lone Rabbinic Comparison

The Jewish tradition, to which I now turn, appears to have had no use at all for the two books metaphor.[31] In fact, most people would be hard-pressed to point to a single traditional rabbinic text in which the study of Torah and the study of nature are at all compared. There are, of course, brief, if potent moments of deliberate blurring of boundaries between science and religion as in the works of Maimonides and some of his immediate followers, and those of Abraham 'Ibn Ezra, but even there one finds nothing even remotely akin to a two books analogy. I have come across only one rabbinical source in which science and Torah-study are explicitly and favorably compared. It does so, however, by thoroughly reversing the two Christian formulae we have discussed.

The text was written during the early 1860's by the famed head of the yeshivah of Volozhin, R. Naftali Tzvi Yehudah Berlin, better known as the NaTZYB (pronounced *Natziv*) of Volozhin. I am referring to the opening lines of Section 5 of *Kidmat ha-Emek*, the NaTZYB's introductory essay to his commentary to the Pentateuch, *Ha-Amek Davar*, in which he explains what motivated him to compose yet another Torah commentary.

> It is easy to understand that just as it is not possible for the wise student of nature ever to boast knowledge of all of nature's secrets..., and just as there is no guarantee that what his investigations do accomplish will not be invalidated in this generation or the next by colleagues who elect to study the same things differently, so it is not possible for the student of the Torah ever to claim that he has attended to each and every point that claims attention, and even that which he does explain—there is never proof that he has ascertained the truth of the Torah! Still, we must do what we can.[32]

---

[31] As Hava Tirosh-Samuelson perceptively notes "In the final analysis, for the Jews there was only one Book—the Torah—whose esoteric dimension constitutes the divine paradigm by which the world was created and whose exoteric linguistic signs provide the interpretative keys to the decoding of God's mysteries, both created and uncreated." Tirosh-Samuelson 1997, 561. I thank an anonymous referee for drawing my attention to this important and perceptive paper in which several themes developed in the present essay are anticipated. See, for instance, her brief account of Bacon and Descartes's philosophies of science and how they "were united in Newton" (536, n. 22).

[32] ונקל להבין, כי כמו שאי אפשר לחכם בסתרי הטבע להתברך בלבבו כי עמד על כל סתרי הטבע. [...] ולא עוד אלא אפילו מה שהעלה בחקירתו, אם אין לו בחינה ברורה כי כך הוא, יכול להיות שיבוא רעהו או בדור שאחריו וחקרו באופן אחר ויסתור בנין קודמו, כך אי אפשר להתברך לחוקר בטבע התורה ולעמוד על כל השינויים והנדרש שום לב ולומר כי עמד על כולן, וגם מה שביאר אין הוכחה שכיון לאמיתה של תורה. ומכל מקום עלינו לעשות מה שבידינו. And see above p. 66.

To recall, both early modern Christian versions of the two books metaphor argue from the proven ability of suitably disposed readers to fathom the true meaning of Scripture to a promise of similar success to students of nature analogously disposed. Rabbi Berlin's argument commences from exactly the same analogy, but runs in exactly the opposite direction, and with the opposite conclusion. Students of Torah, he argues, resemble students of nature (and not the other way around) not in the confidence they are justified to feel about their researches, but, on the contrary, in being persistently *distrustful* of the fruits of their efforts. This is because, just like scientists, students of Torah can never know if the understanding they have achieved is conclusive. Interpretations, convincing as they may be, will always be open to further questioning and to the challenge of alternative readings. More importantly, interpretations can never be proven true. The study of Torah resembles science, argues Rabbi Berlin, in that both undertakings lack criteria for determining either conclusiveness or truth!

According to Rabbi Berlin, Torah study and science are analogous because of the thoroughly pervading constructive skepticism that characterizes both undertakings. Unlike both early modern Christian accounts, for him, what religion and science have in common is a persistent uncertainty regarding the truth of their findings. Just as in science, so in the study of Torah, one can never boast knowledge of the truth, even locally. Torah is the revealed Word of God, of course, but interpreting Torah is always the work of human exegetes, and can, therefore, never transcend the boundaries of inconclusive human conjecture.

Rabbi Berlin does not endorse science as a vocation of any religious merit for Jews,[33] but neither does he deride science, as much religious rhetoric often does. On the contrary, his attitude toward science is clearly favorable. And since he obviously draws the analogy between science and religious study with a view to promoting the latter, his text strongly suggests that in his opinion the perpetual uncertainty he associates with science and Torah study is not viewed as an impediment to their progress, but as the very reason for their great vitality and growth! Humbly aware of their own limitations, scientists and students of Torah become equally

---

[33] Though, surprisingly perhaps, he does view science as a religiously worthy vocation for gentiles. Toward the end of Section 4 of *Kidmat ha-Emek*, a few lines before the paragraph quoted above, and again in his commentary to Deut. 32:40, he describes the work of deciphering the two books (though he doesn't employ the term) as a sort of religiously meaningful division of labor between the Jewish people and "the nations of the world."

aware of those of their predecessors. Because they can never vouch for the truth and completeness of their researchers, all they can do is to criticize former efforts as best they can, with a view to presenting their own better alternatives, which in turn they genuinely hope their students will treat in similar fashion. Science and Torah-study, according to Rabbi Berlin, are both on-going, open-ended exercises in epistemic humility; consisting of creative, yet, at the same time, humbly self-doubting, cycles of keen trial and imaginative error.

To many readers all of this must sound deliciously anarchic coming from so very orthodox an Eastern European rabbi, but in fact it is far less surprising, and certainly far less radical than it might seem, as those familiar with the world of Jewish learning will surely attest. What certainly *is* intriguing is the so very Popperian view of science Rabbi Berlin seems to have internalized so long before Popper.[34] But as far as rabbinical attitudes to biblical exegesis go, Rabbi Berlin's description rings not only true, but is commonplace.

The idea of there being a single, true and humanly discernible reading of the biblical text is as foreign to the Jewish exegetical tradition as it was central to early modern Christianity. From the first tannaitic midrashic compilations of the third century of the common era, the rabbinical canon presents throughout a clamorous, jarring and unresolved polyphony of conflicting voices to which each new generation of learners is invited to add its own—and most often does. Students of the Torah strive to perfect their readings by keenly and critically engaging others, but it is the din of the study-hall rather than any one victorious voice that the canon ends up preserving. The compilers and editors of this literature show no interest at all in adjudicating or deciding the interpretative disputes, only in recording them in all their richness and diversity. Here and there the Talmudic literature hints that this interpretative plurality is not the result of human shortsightedness but is of the very nature of the enterprise. In a discussion of *Judges* 19, in tractate *Gittin* of the Babylonian Talmud, folio 6b,[35] R. Evyatar and R. Yonatan are told to have disagreed as to the reasons for the gruesome fate suffered by the concubine of Bethlehem. They agreed that

[34] As far as we know the Berlin text dates from the early 1860s. Karl Popper published the first systematic analysis of science along these lines only in 1934. On the philosophical and scientific background to the formation of Popper's philosophy see Hacohen 2000, particularly chapters four and six. On the growing awareness in late nineteenth-century discourse of science to the essential open-ended, conjectural nature of all scientific theory, see Herbert 2001.

[35] Henceforth references to the Babylonian Talmud will be of the form: Bavli, *Gittin*, 6b.

her Levite master had needlessly overreacted to some minor negligence on her part, but differed as to its precise nature. Elijah then appeared to one of them in a dream. "What is The Holy One, blessed be He, doing at this time?", he inquired of the sage. "He is busy [studying] (the episode of) the concubine in Gibeah" answered the late prophet. "And what does *He* say (about it)?" urged the sage. He says, replied Elijah, that "Evyatar my son says so and Yonatan my son says differently." "God forbid!," exclaimed the sage, "that there should be doubt even before the Heavenly One!"—to this Elijah replied: *eilu va-eilu divrei Elohim Hayim*—"both are the words of the living God"![36]

The story of Noah's ark offers another good example of the type of exegetical pluralism of which I speak. In this case, however, the debate is about much more than discerning the facts of the matter. It is recorded briefly in *Bavli, Sanhedrin* 108b and related fully in *Bereshit Rabbah* (the amoraic midrashic commentary to Genesis).[37] In Genesis 6:18 we read: "and thou shalt come into the ark, thou, and thy sons, and thy wife, and thy son's wives with thee." Sensitive to the slightest textual irregularity, the midrash takes note of the rather artificial separation of husbands and wives in the phrasing of God's instructions to Noah, concluding that, apart from everything else, they included a prohibition on cohabitation: "whence Noah entered the ark copulation was forbidden," hence, 'thou and thy sons' to themselves, 'and thy wife and thy son's wives' to themselves." Read along these lines, Gen. 7:5–7 is taken to imply that Noah did as he was told: "So Noah did according to all that God commanded him … And Noah went in, and his sons, and his wife, and his sons' wives with him into the ark." In keeping with this reading, upon leaving the ark a year or so later, family life, it seems, was allowed to return to normal, for God's instructions no longer implied the same separation of the sexes: "Go out of the ark, thou, and thy wife, and thy sons, and thy son's wives with thee" (Gen. 8:16).[38] This time, though, Noah appears to have decided not to comply, choosing, seemingly on his own initiative, in the words of the midrash "to extend the commandment": "And Noah went out, and his sons, and his wife, and his sons' wives with him" (Gen. 8:18). And it is here that the following debate between

---

36 מאי קא עביד הקב"ה? א"ל: עסיק בפילגש בגבעה, ומאי קאמר? אמר ליה: אביתר בני כך הוא אומר, יונתן בני כך הוא אומר, א"ל: ח"ו, ומי איכא ספיקא קמי שמיא? א"ל: אלו ואלו דברי אלהים חיים.

37 Midrash *Bereshit Rabbah* 1996, 286, 328.

38 ר' יהודה' בר' סימון ור' חנן בשם ר' שמואל בר רב יצחק נח כיון שניכנס בתיבה נאסר לו פריה ורביה ה"ה ובאת אל התיבה אתה ובניך (בראשית ו יח) לעצמך ואשתך ונשי בניך (שם) לעצמן, כיון שיצא התיר לו ה"ה צא מן התיבה אתה ואשתך וגו'.

R. Yehudah and R. Nehemiah is recorded apropos Gen. 9:8–9: "And God spoke to Noah, and to his sons with him, saying, And behold, I establish my covenant with you etc." The former is of the opinion that since Noah "transgressed the commandment" he was disgraced, and is no longer addressed personally by God. The latter disagrees, concluding, by contrast, that "since Noah extended the commandment and elected to conduct himself in holiness, he and his sons were rewarded by God's word."[39]

The controversy is fundamental. Does the Torah teach us, as R. Nehemiah would have it, that one achieves sanctification by suppressing the flesh, or rather, as R. Yehudah opined, by appropriately acknowledging and fulfilling one's bodily needs? Is Noah to be condemned or praised in 'going beyond the line of duty' in this respect? The two profoundly conflicting religious anthropologies and subsequent readings of Scripture are sharply stated, and played off against each other, but no attempt is made, here or elsewhere, to decide the issue.

To the best of my knowledge, the Talmudic literature nowhere complains about the interpretative under-determination of Scripture. On the contrary, it seems, as a rule, to rejoice in the spirited, creative, on-going, free-for-all that it perpetuates. Scripture seems to function for Jews less as a source of religious truth and value than as a divine invitation—better, perhaps, divine 'challenge'—to try and work them out for themselves. The study of the Torah is as personal and as unpoliced an enterprise as its Lutheran counterpart, but differs from it in not advocating or practicing literalism. Students of the Torah are encouraged to bring their theologies, value systems and religious sensibilities to the text, yet unlike Catholicism, the rabbinical tradition does not insist that these be doctrinal. Most importantly, with regard to biblical exegesis, the Jewish tradition differs from both early modern Christian approaches in not presupposing a single point of interpretative convergence. It solemnly recognizes the *Torah she-bikhtav*—the Written Torah—the Holy Text itself—as sacred and binding, without recognizing any particular *reading of it* as sacred or binding! In fact one gets the impression that the Bible's very sanctity is attested to by the lavish abundance of conflicting readings it can, and is in fact given.

And yet, it is not as if "anything goes" in the study of the Torah, to quote Paul Feyerabend's radical slogan.[40] The rabbis take their Scripture with the

39 ויאמר אלהים אל נח ואל בניו וגו' ואני הנני מקים את בריתי וגו' ר' יהודה א' לפי שעבר על הציווי לפיכך נתבזה, ר' נחמיה א' הוסיף על הציווי ונהג בקדושה לפיכך זכה הוא ובניו לדיבר ויאמר אלהים אל נח ואל בניו וגו'.

40 Feyerabend 1977.

utmost seriousness, paying close attention to every word, and every unusual or seemingly redundant turn of phrase. The divine text is not allowed to mince its words, and certainly not to waste them.[41] None the less, much leeway remains, and among the very many possibilities, none are deemed obligatory. Rabbi Berlin's fascinating description of Torah-study as a constructively skeptical and self-doubting, critical engagement with tradition (akin to modern science), is far less a radical battle-cry than an apt and commonplace observation. The Bible's narratives, its portrayals of God, of humankind, of Israel, and of the nations; its accounts of creation, of the end of days, and of the world to come; its poetics, prophecy and philosophy, are all energetically discussed, heatedly disputed, yet nowhere adjudicated and never decided. I can think of no figure of standing in the long history of Jewish Biblical commentary who would dispute Rabbi Berlin's characterization.

Halakhic Authority in Dispute

But while this is widely true of the Jewish exegetical tradition, it is also true that Judaism involves more than an effort to interpret Scripture. Judaism is also, if not primarily, a form of life constituted by an intricate set of practices that are governed by a complex system of religious law—halakha; a religion of meticulously regulated works, not merely one of words. And with regard to the legislative, judicial and performative aspects of Jewish law, to its norms of conduct, to what Jews are required to do and to refrain from doing—as opposed to what they choose to think—does one still find the same happy, pluralistic, disputative polyphony? Does Rabbi Berlin's acute portrayal of the Jewish interpretative tradition extend to the normative and the legal? Rabbi Berlin himself did not think it does.[42] A very different consensus, he along with most observant Jews believe, grounds and regiments the realm of religious law and divine commandment, rendering

[41] It is, therefore, not allowed to state the obvious. "*ve-ha-lo din hu*?!" is the exegetical question that often prompts a midrashic move in the early tannaitic compilations—"Is it not obvious?!" meaning: how can the Torah appear to be teaching us something that we could figure out ourselves? The *midrash* always responds to this question by interpreting the text non-literally so that its message be something new that we could not have learned elsewhere. On the important differences between the midrashic schools of R. Ishmael and R. Akiva with regard to such issues see Yadin 2004, esp. ch. 6.

[42] For Rabbi Berlin's very different account of the history of halakhic discourse, see his introductory essay "The Way of Torah" Berlin 1986, Vol. I, 5–12. I discuss the two different models of development employed by Rabbi Berlin in Fisch 1997, 286–8.

it, very differently from biblical exegesis, profoundly monolithic. And if that is so, then despite its exegetical skeptical pluralism, Judaism too would easily fall prey to the critique of modernity of which we have spoken.

Unlike biblical exegesis, religious law and divine commandment are binding, deriving their normative mandate from three main sources of religious authority: the halakha—the accumulating body of Jewish law, the halakhic authorities, namely, the courts and sages whose job it is to make and state the law, and The Almighty himself, of course, whenever He chooses to make known his will directly. Here, in the realm of religious performance, as opposed to that of biblical exegesis, many believe that religiosity for Jews is strictly a matter of devout, uncritical submission, in which one does not challenge authority or tradition, but undertakes piously to submit to their ruling. The covenant of Sinai is one of solemn obedience, they insist, summarized forcefully by the phrase *na'ase ve-nishma* (Ex. 24:7)—first we undertake to do, and then, only then, do we hearken and take to heart.

Moreover, many Jews feel that it is the submissive unity created by uncritical compliance with the law that explains and enables Judaism's unique interpretative pluralism. Hobbes was right, they imply, a religion can simply not function as a religion if its members are allowed—nay encouraged!—to interpret its sacred texts as each sees fit. However, rather than freely submit to the Leviathan's reading of Scripture, as Hobbes insisted, Judaism leaves Scripture free, as it were, and has its own 'Leviathan' firmly dictate the law.[43]

I do not deny that such a view exists within rabbinical Judaism. On the contrary, it is an authentic voice of the tradition. But I would strongly oppose the claim that it is the tradition's *only* voice. Contrary to the broad consensus surrounding Berlin's characterization of biblical exegesis, no consensus exists within the formative canon of Judaism with respect to halakha. On the question of obedience and submission to the three sources of halakhic authority, the tradition, I shall argue, is deeply and intriguingly divided—certainly at the formative level of the rabbinic literature of late antiquity, on which I shall be focusing. My claim shall be that at this most fundamental level, the Jewish sources are in profound disagreement as to the very meaning of religious commitment and obligation! Judaism, one

[43] For an enlightening comparison of Hobbes's approach to exegetical authority to that of the Jewish tradition see H. Fisch 1993.

could say, is a religion constituted by a foundational dispute about what religion should be.

Take for example the following dispute with respect to Deut. (17:11). The verse states that:

> According to the sentence of the Torah which they shall teach thee, and according to the judgment which they shall tell thee, thou shalt do; thou shalt not deviate from the sentence which they shall tell thee, to the right hand or to the left.

Appalled, as Thomas Hobbes would one day be, of the possibility of a religious society whose leadership did not enjoy absolute authority, Nahmanides, the great thirteenth-century halakhist and exegete, reads the verse as a firm injunction to submit to the court's authority:

> Even if you think in your heart that [the court is] mistaken, and even if the matter is as obvious to you as the difference you discern between right and left, you are to follow their command! Do not say, how can I eat this totally unkosher animal fat, or how can I kill this innocent man, but say thus the Lord commanded me, who commands all the commandments that I fulfill, that I will fulfill all His commandments according to everything they, who stand before Him in the place he chose teach me, and according to the meaning they understand in it He gave me the Torah, even if they err … And the Torah determines that we should listen to everything the high court that stands before God in the place He chose says regarding the interpretation of Torah … Because according to their understanding He gave us the Torah, even if it seems to us as if they confused right and left.[44]

It is hard to imagine a clearer and more vigorous statement of the religiously required obedient surrender to the institutions of halakha.[45] However,

---

[44] Nahmanides to Deut. 17:11.
אפילו תחשוב בלבך שהם טועים, והדבר פשוט בעיניך כאשר אתה יודע בין ימינך לשמאלך, תעשה כמצותם, ואל תאמר איך אוכל החלב הגמור הזה או אהרוג האיש הנקי הזה, אבל תאמר כך צוה אותי האדון המצוה על המצות שאעשה בכל מצותיו ככל אשר יורוני העומדים לפניו במקום אשר יבחר ועל משמעות דעתם נתן לי התורה אפילו יטעו...וחתך לנו הכתוב הדין, שנשמע לבית דין הגדול העומד לפני השם במקום אשר יבחר בכל מה שיאמרו לנו בפירוש התורה...כי על הדעת שלהם הוא נותן לנו התורה, אפילו יהיה בעיניך כמחליף הימין בשמאל.

[45] But note that granting this absolute authority to the courts *even if they err* means granting them absolute freedom to determine the law according to their understanding even if it deviates from the original intention of the Scriptures. From the phrase, "even if they err," we learn that according to Nahmanides the interpretation of the sages does not establish the meaning of the verse, but, rather, seeks to understand its intention. See Sagi 1993, esp. 130–31. My concern here, of course, is less with the interpretative freedom Nahmanides grants the courts, than with his demand for submissive, uncritical acceptance of their rulings.

Nahmanides knew quite well that in adopting the *Sifre*'s interpretation of "thou shalt not deviate,"[46] he was not stating the obvious, but making a conscious choice between two contradictory alternatives found in rabbinic literature—between the *Sifre*'s rendition of the verse and the very different reading found in tractate *Horayot* of the Jerusalem Talmud:

> Can it be the case that if [the court] instructs you that right is left and left is right, that you should obey them?! Scripture therefore teaches us: 'to the right or to the left' (Deut. 17:11): that [is to say, that only if] they say to you, right is right and left is left, [should you obey].[47]

According to the *Yerushalmi*, if one prudently believes that the court, even the High Court of Jerusalem, has ruled mistakenly, one's religious obligation is not to comply! And it is hard to think of a clearer or more vigorous expression of a critical, nonsubmissive religious attitude towards the institutions of halakha![48]

Welcome Disobedience

The *Sifre* and the *Yerushalmi*'s very different renditions of the same verse clearly illustrates how the dispute of which I speak plays itself out with regard to the institutions of halakha. Let us now see how it manifests itself with regard to the ultimate source of religious authority, namely, God Himself. To this end let us look briefly at two very different midrashic renditions of the same rather surprising biblical moment.[49] In the course of his long farewell speech, recorded in Deuteronomy, Moses recalls how upon nearing the Holy Land he was instructed by God:

> Saying: "Rise, take your journey, and pass over the brook Arnon; behold, I have given into your hand Sihon the Amorite, king of Heshbon, and his land; begin to possess it, and provoke war with him. For this day will I begin to put the dread of you and the fear of you upon the nations that are under

46 The third century, tannaitic midrash halakha *Sifre on Deuteronomy* interprets the verse as instructing that "even if in your eyes [they are ruling that] right is left and left is right, you must obey them" (*Sifre on Deuteronomy* 1993, 207).

47 Yerushalmi, *Horayot* 45d.
יכול אם יאמרו לך על ימין שהיא שמאל ועל שמאל שהיא ימין תשמע להם? תלמוד לומר:
ללכת ימין ושמאל—[רק] שיאמרו לך על ימין שהוא ימין ועל שמאל שהיא שמאל.

48 For a detailed discussion of the dispute, see Sagi 2007, ch. 12, Blidstein 1997, and Lorberbaum 1995.

49 For a discussion of these midrashim along lines similar to the thesis proposed here, see Schremer 2001, esp. 759–63.

> the whole heaven, who shall hear the report of you, and shall tremble, and be in anguish because of you." [But] I sent messengers out of the wilderness of Kedemoth to Sihon king of Heshbon with words of peace, saying: "Let me pass through your land; I will go along by the high way, I will neither turn to the right hand nor to the left . . ." (Deut. 2:24–28)

Moses, it seems, chose knowingly to disobey God's explicit command, deciding on his own accord to seek peace with the Amorites rather than provoke war as instructed. *Devarim Rabbah* (the amoraic midrashic commentary to Deuteronomy) explains his insubordination thus:

> Although the Holy One, blessed be He, instructed Moses 'Begin to possess', he did not so do, but rather 'sent messengers;' Although He told him 'Make war', he sought peace! [He did so, however,] because it is written in the Torah: 'When thou comest near to a city to fight against it, then proclaim peace unto it' (Deut. 20:10), that is why he sent words of peace to Sihon, for it is written 'out of the wilderness of Kedemoth', on account of words of the Torah that come before those of the Almighty. Therefore it says 'words of peace.'[50]

This, I believe, is about as submissive a reading the Deuteronomy verses can be given without doing them real violence. Moses is described as apologizing piously for not heeding to the Lord's instruction to provoke war, arguing, that he is not at liberty to defy the Torah's explicit commandment always to propose peace before attacking. In case of contradiction, he explains, the words of the Torah are to 'come before' those of God.[51] Even within a thoroughly submissive religious framework, questions will always arise regarding possible or actual conflicts between its various sources of authority. According to this midrashic reading, the question Moses faced and answered was to which of the conflicting sources of authority should he properly submit, the Torah that instructs always to first seek peace or the Almighty who instructed to provoke war. Moses opts for the former, but questions neither.

---

50 *Devarim Rabbah* 1964, 29–30.
אף על פי שאמר לו הקב"ה למשה החל רש, משה לא עשה כן, אלא ואשלח מלאכים. אע"פ שאמר לו לך עשה מלחמה, ביקש שלום. שכן כתוב בתורה כי תקרב אל עיר להלחם עליה וקראת אליה לשלום (דברים כ, י), לכן שלח אל סיחון דברי שלום, שנאמר ואשלח מלאכים ממדבר קדמות (דברים ב, כו), מדברי תורה שקדמו לדבריו של הקב"ה, לכך נאמר "דברי שלום".
The last sentence of the paragraph clearly appears corrupted in its present form. "From words that came before the Torah, from words of God. Therefore it says 'words of peace.'" It should be corrected as proposed as a play on the word *kedemoth* read as a derivative of the Hebrew verb *k.d.m.*—precedes, comes before.

51 Compare Schremer 2001, 760–61.

Bamidbar Rabbah (the amoraic midrashic commentary to Numbers) 19:20 tells a different story entirely, however.

> This is one of three things said by Moses to the Holy One, blessed be He, to which the latter replied: 'You have taught Me something!'…when the Holy One, blessed be He, said to him: 'Make war with Sihon. Even though he does not seek to interfere with you, you must provoke war with him'…Moses did not do so, but, as it is written further down, 'sent messengers […with words of peace].' The Holy One, blessed be He, said to him 'By your life! You have spoken well! You have taught Me something! And I shall thereby cancel My words and adopt yours'; hence it is written: 'When thou comest near to a city to fight against it, then proclaim peace unto it' (Deut. 20:10).[52]

Unlike the submissive Moses of *Devarim Rabba*, in a bind to decide between the conflicting instructions of the Torah and its Author, Moses is here forcefully portrayed as having challenged God's decree by refusing to provoke war with Emorites on moral grounds. God, in response, is said not to have demanded obedience, but to have heeded to Moses' objection and to have introduced the required changes in the Written Torah! According to the first midrashic rendition of the story, Moses disobeyed God's explicit command because the Torah instructs differently; according to the second midrash, the Torah ends up instructing differently because Moses disobeyed!

The other two examples cited by the midrash are equally dramatic. According to one of them, upon hearing the Almighty declare at Sinai "for I the Lord thy God am a jealous God, punishing the iniquity of the fathers upon the children unto the third and fourth generation," Moses is said to have objected: "Sovereign of the universe! Many are the wicked who have begotten righteous sons. Shall the latter bear some of the iniquities of their fathers?…Is it proper that the righteous should be punished for the iniquity of their fathers?" To which God is reported to have replied as above: "You have taught Me something! By your life! I shall cancel My words and adopt yours!"[53] For indeed it is (thereafter) written "Fathers shall not be put to death for children, neither shall children be put to death for fathers,

---

[52] Numbers *Rabbah* (Vilna), 19:33.
זה אחד מג' דברים שאמר משה לפני הקב"ה ו[והקב"ה] מר לו למדתני...כשאמר לו הקב"ה עשה מלחמה עם סיחון אפילו הוא אינו מבקש לעשות עמך את תתגר בו מלחמה, שנאמר (דברים ב) קומו סעו ועברו את נחל ארנון, ומשה לא עשה כן אלא מה כתיב למעלה ואשלח מלאכים. אמר לו הקב"ה חייך שאני מבטל דברי ומקיים דבריך, שנאמר (דברים כ) כי תקרב אל עיר להלחם עליה וקראת אליה לשלום.

[53] Ibid.

every man shall be put to death for his own sin." (Deut. 24:16)[54] As this midrash has it, the principle of individual liability was introduced into the Torah by God as an afterthought following Moses' forceful criticism of the Ten Commandments!

In both cases God is astonishingly described as having produced a morally deficient early draft of the Torah, Moses is portrayed as having refused to comply, and having challenged it on moral grounds, and the Written Torah, the very Word of God, is subsequently described as having been happily revised by virtue of God accepting Moses' superior moral judgment!

Confrontational Theology

The theology at work here, and especially the notion of religiously appropriate conduct implied by these passages differs radically from the kind of pious submissive dogmatism one normally associates with religion. The God here portrayed is so very different from the all-powerful, all-knowing and especially perfectly good deity to whom we have become so accustomed in virtually all God-centered religious discourse. The God of this midrash and its likes is a covenanting God, but the covenant his human partners are invited to enter, is not one of religious obedience and surrender. It is better described as 'a covenant of constructive confrontation,' especially moral confrontation; a sacred partnership dedicated by both partners to the betterment of the world, the betterment of the Torah and the betterment of the self. God's need for the critical input of his human partners is as real as theirs should be for his. It is, therefore, a partnership of mutual learning and improvement. God makes known his imperfection by initiating real engagement, by learning and by admitting his mistakes. And in doing so urges humans, by example, to openly confront each other and acknowledge theirs.

Confronting the Almighty can be terrifying, of course, and taking moral responsibility is never easy. The temptation of pious surrender is ever-present. The opening sections of the *Mekhilta*, the tannaitic midrashic

---

54 Ibid.
אמר משה רבש"ע כמה רשעים הולידו צדיקים יהיו נוטלין מעונות אביהם תרח עובד צלמים ואברהם בנו צדיק וכן חזקיה צדיק ואחז אביו רשע וכן יאשיה צדיק ואמון אביו רשע וכן נאה שיהו הצדיקים לוקין בעון אביהם אמר לו הקב"ה למדתני, חייך שאני מבטל דברי ומקיים דבריך שנאמר (דברים כד) לא יומתו אבות על בנים ובנים לא יומתו על אבות.

commentary to Exodus, offers another powerful example of this approach.[55] There are three classes of prophets, states the *Mekhilta*: those who will plead both the cause of the father against the son *and* that of the son against the father, and those who will only plead the former or only plead the latter, but not both. The perfect prophet is described as a Janus-faced critic who chastises the people in the name of God, but also challenges God's judgment on behalf of the people. Jeremiah is approvingly depicted as having been such a prophet—"We have transgressed and have rebelled," (Lam. 3:42) he rebukes the people, but "thou hast not pardoned!", he is read as complaining against God. Elijah, on the other hand, is portrayed as having avidly reproached the people for their idolatrous ways (citing the spectacular scene on Mt. Carmel), but of not ever questioning God's judgment or actions. Elijah's replacement by Elisha is described dramatically as an act of impatient divine dismissal owing to God's reluctance to employ prophets who like Elijah refrain from speaking back! "We have no use for your prophecy," God is said to have stormed, "go anoint Elisha in thy stead!" This rabbinical voice has little patience with religious yesmanship.[56]

Two objections might be raised at this point, however. First, one might object to the attempt to infer a general perception of religiosity from depictions such as these that clearly concern God's dealings with, and expectations from, such mythical, larger-than-life, saintly figures of old as Moses, Jeremiah and Elijah who obviously enjoy privileged status in Jewish religious imagination. Second, even if such an inference could be made, it could still be claimed that in halakha one still finds the truly submissive

---

55 *Mekhilta D'Rabbi Ismael* 1997 Tractate *Pescha, Bo*, 1, 4.

56 נמצאת אתה אומר שלשה בנים הם אחד תבע כבוד האב וכבוד הבן. ואחד תבע כבוד האב ולא כבוד הבן. ואחד תבע כבוד הבן ולא בכבוד האב. ירמיה תבע כבוד האב וכבוד הבן שנאמר נחנו פשענו ומרינו אתה לא סלחת (איכה ג מב) לכך נכפלה נבואתו שנאמר ועוד נוסף עליהם דברים (ירמיה לו לב) אליהו תבע כבוד האב ולא כבוד הבן שנאמר קנא קנאתי ליי' אלהי צבאות וגו' (מלכים א' יט יד) ומה נאמר שם ויאמר ה' אליו לך שוב לדרכך מדברה דמשק וגומר ואת יהוא בן נמשי תמשח למלך על ישראל ואת אלישע בן שפט תמשח לנביא תחתיך שאין תלמוד לומר לנביא תחתיך אלא שאי איפשי בנבואתך. יונה תבע כבוד הבן ולא כבוד האב [שנאמר ויקם יונה לברוח וגו' מה כתיב] ויהי דבר יי' אל יונה שנית לאמר (יונה ג א) שנית נדבר עמו לא שלישית.

Schremer 2001 rightly describes these texts as premising a notion of religiosity that is not characterized by passivity and obedience, but rather "stands up against the tradition, criticizing, and even rejecting it, whenever it seems unacceptable." Schremer rightly points out the essential confrontational nature of this form of religiosity, but I believe wrongly identifies its critical thrust as directed exclusively against the tradition. Texts such as those examined here clearly attest to a confrontational attitude toward all of Judaism's sources of religious authority—tradition, but not only tradition, included. As we have seen, apart from tradition, the sources of authority toward which such a stance is valorized include the institutions of halakha, and, astonishing as it may sound, God himself!

element of the Jewish faith. Although the tradition contains brave dramatic, yet essentially narrative moments of confrontation with the divine, the law, many would argue, remains the law whatever. When it comes to halakha proper, it might be objected, even this kind of dramatic conflict with the Almighty gives way to full submissive compliance.

In the Face of Disaster

The following juxtaposition of texts proves both objections to be groundless. Again, the voice of religious submission speaks loud and clear, but it does not speak alone. Self-consciously opposing it one finds clear expression of an avid confrontational approach even to halakha, an approach that asserts itself by extrapolating explicitly from tales of biblical heroes to norms of every-day religious conduct and expression. The dispute in question concerns the religious appropriateness of prayers of anguish. The first text, Ch. 9 of Mishna, *Berakhot,* contains one of the keenest expressions of submissive religiosity to be found in the entire Talmudic literature.

Tractate *Berakhot* details the laws and regulations concerning obligatory prayer. Its ninth and final chapter is devoted to obligatory expressions of awe, wonder, and gratitude in acknowledgment of God's awesome presence and good work in the world. It commences with the obvious:

> If one sees a place where miracles have been wrought for Israel, he should say: "Blessed [art thou, O Lord our God, King of the universe] who wrought miracles for our ancestors in this place." On seeing a place from which idolatry has been extirpated, he should say: "Blessed [art thou . . .] who extirpated idolatry from our land."
>
> [On witnessing] shooting stars, earthquakes, thunderclaps, storms and lightning one should say: "Blessed [art thou . . .] whose strength and might fill the world." On seeing mountains, hills, seas, rivers and deserts he should say: "Blessed [art thou . . .] who wrought creation."[57]

The Mishna's submissiveness becomes cruelly apparent when it moves from the wonderful and the awe-inspiring to the catastrophic.

---

[57] Mishna, Berakhot 9:1–2.
הרואה מקום שנעשו בו ניסים לישראל אומר ברוך שעשה ניסים לאבותינו במקום הזה מקום שנעקרה ממנו עבודת כוכבים אומר ברוך שעקר עבודת כוכבים מארצנו:
על הזיקין ועל הזועות ועל הברקים ועל הרעמים ועל הרוחות אומר ברוך שכחו וגבורתו מלא עולם על ההרים ועל הגבעות ועל הימים ועל הנהרות ועל המדברות אומר ברוך עושה מעשה בראשית.

> For rain and for good tidings one says: "Blessed [art thou...] who is good and bestows good."
>
> For bad tidings one says: "Blessed [art thou, O Lord our God, King of the universe] who is the true judge." ... Over the dreadful a blessing is said similar to that over the good, and over the good, a blessing is said similar to that over dreadful.[58]

Standing before God, the Mishna rules, the appropriate religious response to suffering is one of capitulation, of total acknowledgment of his supreme judgment. Pain and suffering exist in the world, the Mishna implies, but there exists no evil, no mischief, no oversight. People may scheme to do evil, natural catastrophes may hit, accidents are liable to happen, but their outcomes, dreadful as they may be, unfair as they may seem, are the doings of "the true judge," the perfect workings of divine justice. Standing before God in response to tragedy, one should, therefore, not merely resign oneself to one's fate, but be thankful for it—always. The Mishna declares petitional prayers of anguish to be religiously inappropriate (as the taking of God's name in vain) if they are uttered retrospectively in tortured response to past occurrences. It is not because God is powerless to intervene, but because in a world overseen by a perfectly just, personally intervening deity, all that occurs is deemed to be just by the very fact that it has occurred! Prayers of anxious appeal are only appropriate prospectively, in fearful anticipation of what the future might hold, never, but never after the event.

> For to cry out (in prayer) over the past is to utter a prayer in vain. For instance, If a man's wife is with child and he prays: "May it be Thy will [O God] to grant that my wife bears a male child" this is a prayer in vain. [Similarly] if a person is coming home from a journey and he hears cries of distress in the city, and he prays: "May it be Thy will [O God] to grant that this is not in my household," this is a prayer in vain.
>
> For what has past, one should at all times give thanks, and cry out [only] for what is yet to come.... As it says: "And thou shalt love the Lord thy God with all thy heart ... and with all thy might" ... 'with all thy might'—for each and every measure that He might measure out for you, you shall thank Him mightily.[59]

---

58 Ibid 9:2.
על הגשמים ועל הבשורות הטובות אומר ברוך הטוב והמטיב ועל שמועות רעות אומר ברוך דיין האמת:
...מברך על הרעה מעין הטובה ועל הטובה מעין הרעה.

59 Ibid 9:3–5.
הצועק לשעבר הרי זו תפלת שוא. כיצד? היתה אשתו מעוברת ואמר יהי רצון שתלד אשתי זכר הרי זו תפלת שוא, היה בא בדרך ושמע קול צוחה בעיר ואמר יהי רצון שלא יהיו אלו בני ביתי הרי זו תפלת שוא:
...ונותן הודאה לשעבר וצועק לעתיד לבא:

The Mishna's theology is the theology of Job's submissive friends. It is one of pious surrender. And just like Job's four friends, neither does the Mishna go unchallenged. In its commentary to Genesis 30, the *Tanhuma*, the late fifth- or early-sixth-century Palestinian midrashic compilation, confronts both the Mishna's ruling and implied theology head-on. It does so apropos the story of the birth of Leah's daughter, Dina, and in the most forceful terms. The *Tanhuma* openly questions the Mishna's ruling on the religious appropriateness prayers of anguish:

> "And afterwards she [Leah] bore a daughter" (Gen. 30:21). May our Rabbi teach us: If a man's wife is with child is one allowed to pray "May it be Thy will that my wife bears a male child"—for so teaches [the Mishna]: "If a man's wife is with child and he prays: 'May it be Thy will [O God] to grant that my wife bears a male child' this is a prayer in vain"?
>
> R. Huna in the name of R. Yossi says, despite the fact that [the Mishna] teaches that "If a man's wife is with child and he prays: 'May it be Thy will [O God] to grant that my wife bears a male child' this is a prayer in vain," [the law] is not so. Rather, one may pray for a son even as she is giving birth, because it is not beyond God's power to change females to males and males to females.[60]

It is very rare indeed to find an essentially exegetical source such as this explicitly contesting a major tannaitic ruling. The *Tanhuma* is not suggesting that the Mishna has been misunderstood, but boldly asserts that it is wrong! And it does so not by piously appealing to a higher halakhic authority, but by contesting the Mishna's theological basis. The question, the *Tanhuma* dramatically implies, is not a technical one, but pertains to the very foundation of the religion. "For so says Jeremiah," the *Tanhuma* continues:

> "Then I went down to the potter's house, and, behold, he was at work on the wheels. And the vessel that he was making of clay was spoiled in the hand of the potter: so he made it again into another vessel, as it seemed good to the potter to make it. Then the word of the Lord came to me, saying, O house of Israel, cannot I do with you as this potter? said the Lord. Behold, as the clay in the potter's hand, so are you in My hand, O house of Israel." (Jer. 18:3–6). Says the Almighty to Jeremiah: if a potter can break a cruse

---

...שנא' (דברים ו) ואהבת את ה' אלהיך בכל לבבך ובכל נפשך ובכל מאדך...בכל מאדך בכל מדה ומדה שהוא מודד לך הוי מודה לו במאד מאד...

60 ילמדנו רבינו אם היתה אשתו מעוברת מהו שיתפלל יה"ר שתלד אשתי זכר, כך שנו רבותינו היתה אשתו מעוברת ואמר יה"ר שתלד אשתי זכר הרי זו תפלת שוא, רב הונא בשם רבי יוסי אומר אע"פ ששנינו היתה אשתו מעוברת ואמר יה"ר שתלד אשתי זכר הרי זו תפלת שוא אינו כן אלא אפילו עד שעה שתשב על המשבר יתפלל שתלד זכר לפי שאין קשה לפני הקב"ה לעשות את הנקבות זכרים ואת הזכרים נקבות.

> after he has made it and make it into something else, can I not do likewise to you O House of Israel!?[61]

Not every state of affairs for which God is responsible comes out well, the *Tanhuma* reads the Jeremiah verses as claiming. They liken him to the imperfect potter whose handiwork is not always up to par. But by citing these verses in the course of discussing prayers of anguish, the *Tanhuma* lends them an additional and radical twist. Jeremiah's potter is self sufficient. He may be imperfect, but he is an effective self-critic who can tell when a vessel comes out right or wrong. Jeremiah's divine potter is his own best critic, as it were. But in drawing on the potter analogy in support of the appropriateness of prayers of anguish and complaint, the *Tanhuma* implies that God is not self-sufficient in this respect, that not only is his handiwork occasionally flawed, but that if the flaws are not pointed out to him, He is liable not to notice them. This, as the subsequent passage makes clear, is the theology that lies at the heart of the *Tanhuma*'s radically and self-consciously different approach to prayer and to the religious covenantal relation in general.

> And similarly in the case of Leah, who after bearing [Jacob] six sons, saw prophetically that twelve tribes were to emerge from Jacob. Having born six already, and pregnant for the seventh time, and with two sons already born to each of the two maidservants, ten had already been born.
>
> Therefore, Leah stood, angrily confronting the Almighty, saying: Lord of the Universe, twelve tribes are to emerge from Jacob, of which I have six, and am pregnant with a seventh, and by means of the maidservants two and two, hence there are ten. If this (unborn) child (I am carrying) is also a son, then my sister Rachel's share will not (even) be that of the maidservants!

At this point the midrash could have easily described God as having scolded Leah in the way He had scolded the laughing Sarah who had doubted his promise of a child (Gen. 18:14). "Do I not know that this would not be fair to Rachel?! Of course you are carrying a Girl!", the midrash could have had God respond. But the *Tanhuma* opts for the far more radical option. God is depicted as having erred in indeed giving Leah a seventh male child. Realizing his mistake, we are told,

> The Almighty immediately responded to her prayer and the boy in her womb was rendered a girl...

61 וכן מפורש ע"י ירמיה (ירמיה יח) וארד בית היוצר והנהו עושה מלאכה על האבנים ונשחת הכלי אשר הוא עושה וגו' ושב ויעשהו כלי אחר כאשר ישר בעיני היוצר לעשות, מה הקב"ה אומר לירמיה הכיוצר הזה לא אוכל לעשות לכם בית ישראל נאם ה'?

> And why did she name her (daughter) Dina? Because Leah the righteous called the Almighty to task.[62]

Here too, as in the stories about Moses mentioned above, God is astonishingly described as being corrected on moral grounds, admitting his mistake, and righting the morally flawed situation for which He was responsible. But unlike the Moses encounters, the *Tanhuma* text conveys two important additional normative points, that prove the two previously mentioned objections groundless. First, the dramatic story of Leah's confrontational prayer is cited explicitly to prove a halakhic point of law that applies to each and every person, namely, that a person should always thus pray if he or she suspects that they or others may not have been dealt with fairly. Unlike the Moses encounters, Leah's confrontation is cited as a model of religious conduct. According to the *Tanhuma*, one is not supposed to suppress prayers of anguish, or to always accept the dreadful as the work of a perfect judge. On the contrary, God, the *Tanhuma* all but states openly, is simply *not* the all-perfect judge, and should not be treated as such by anyone! To act in supreme religious righteousness is to act like Leah, and call God to task whenever one suspects He has ruled wrongly.

Second, and even more importantly for the present context, The *Tanhuma*'s entire move is first and foremost an open, self-conscious challenge to the halakhic tradition as it finds it. The *Tanhuma* confronts the authoritative Mishna's ruling openly. It is a move of the kind I have elsewhere described as 'anti-traditionalist.'[63] In rejecting and contradicting the

---

62 Midrash *Tanhuma*, Vateze 8.
וכן אתה מוצא בלאה אחר שילדה ששה בנים והיתה רואה בנבואה שי"ב שבטים עתידין לעמוד מיעקב וכבר ילדה ששה ונתעברה מן השביעי ומן שתי שפחות שנים הרי עשרה, עמדה לאה והיתה מתרעמת לפני הקב"ה ואמרה רבש״ע י"ב שבטים עתידין לעמוד מיעקב הרי כבר בידי ששה ומעוברת משביעי וביד השפחות שנים הרי עשרה אם זה זכר לא תהא רחל אחותי כאחת מן השפחות, מיד שמע הקב"ה תפלתה ונהפך העובר שבמעיה לנקבה...ותקרא שמה דינה ואחרת אין כתיב כאן אלא ואחר, ולמה קראה לאה דינה לפי שעמדה לאה הצדקת בדין לפני הקב"ה.
The Hebrew subtly makes use of two wordplays. Leah is said to have called her daughter Dina from the word *din*—justice which is the counterpart divine attribute to mercy—Leah's prayer is thus decidedly depicted not as a plea for mercy, which would have implied, even if answered, that in all justice, Dina indeed should have been a male child. Second, Leah is described as "the righteous"—*'tzadeket*—which in Hebrew can be taken to derive from *tzedek*—justice—or *tzidduk*—justification. In the latter case righteousness connotes a pious 'justification of God's judgment' which is the formal halakhic term for the Mishna's obligatory blessing in case of tragedy: "Blessed art thou who is the true judge." In Leah's case the term is meant to connote the exact opposite. That Leah continued to pursue justice in the face of God's unfair judgment!

63 Cf. Fisch 1997, 43–63.

Mishna's ruling as it does, the *Tanhuma* seems clearly to apply to the realm of halakha exactly the same confrontational attitude that many centuries later will be applied so vividly to biblical exegesis by R. Berlin. The *Tanhuma* does not attempt to offer a new reading of the Mishna, but openly challenges, and subsequently overturns what it takes the Mishna to have ruled.

Halakhic Reform

It would be wrong, however, to view the *Tanhuma*'s halakhic antitraditionalism as a late fifth- or early-sixth-century development. The dispute between submissive and confrontational approaches to halakha is foundational to the entire Talmudic project, and is attested to by a variety of tannaitic and allegedly tannaitic texts. To see this let me say a little more about the ways the dispute manifests itself with regard to halakha.

Unlike other forms of religious discourse, halakhic debate requires closure—at least with respect to matters of practical consequence. All agree that halakha has to be decided.[64] Traditionalists and antitraditionalists disagree as to the status of such decisions. The former insist that halakhic rulings are final, and that, therefore, the halakhic tradition—that is to say, the entire body of all past rulings—is at all times binding, not only on practitioners, but even on the halakhic authorities: the courts and the rabbis. They are binding not because they are found to be just or reasonable, but because once a ruling is made it is irrevocable. As one markedly traditionalist tannaitic text—Tosefta, *Sanhedrin* 7:1—puts it, no halakhic institution, not even the Great Sanhedrin—the High Court of Jerusalem itself—has the authority to overturn an existing halakhic tradition. Court discretion states the Tosefta, is wholly limited to halakhic lacunae, to questions for which the tradition at the time has no answer. And once the new question has been decided, and the vote taken, the new matter is considered settled once and for all. Antitraditionalists, by contrast, consider it the religious duty of the halakhic authorities not only to transmit, and to occasionally supplement the system as new questions arise, but to hold past rulings in constant review and to revise them as they see fit.

---

[64] It remains an interesting open question, however, whether halakhic decisions are required to apply nationally, or whether the system can make do with local closure, with different communities deciding differently—as the talmudic literature seems to indicate in certain places. This is a question I cannot address here. For the purposes of the present discussion suffice it to say that halakha has to be decided even if only locally.

A comparison between the Mishna and Tosefta versions of the first chapter of tractate *Eduyot* provides a good example of how the traditionalist-antitraditionalist dispute plays itself out in the early tannaitic literature. Both texts ask why rejected minority opinion should remain on record despite the fact that it is rejected? Both present two answers: one markedly traditionalist and one squarely antitraditionalist. The traditionalist answer is that the reason for keeping records of rejected opinion is to ensure that they remain rejected. (They are remembered, that is, in order to make sure that they remain forgotten!) The antitraditionalist answer is that minority opinions are preserved as a resource for future halakhists who may see fit to overturn their predecessor's majority rulings and revive the opinions rejected. Interestingly, however, the Mishna and Tosefta, decide the matter differently. The Tosefta rules in favor of the traditionalist option by citing it as the majority view and presenting the antitraditionalist answer as a minority, and hence rejected, opinion. The Mishna, by contrast, has it exactly the other way round: ruling in favor of the antitraditionalist answer by citing *it* as the majority view, while presenting the traditionalist option as the rejected minority opinion. The disagreement couldn't be stated more clearly or more explicitly.

Halakhic antitraditionalism is, hence, an exact correlate of the constructive skepticism R. Berlin (exclusively) associates with biblical exegesis. Antitraditionalists treat all halakhic decision-making as tentative and revisable (by the appropriate authorities, of course). Even the great assemblies of the sages of old are liable to have erred or to have failed to see the larger picture, let alone to have not anticipated future changes of circumstance and opinion. And if they are considered fallible, all the more so that we are!

The debate between the 'Houses' of Hillel and Shammai will be my last example. The controversies between the two schools of halakhic thought who, according the Talmudic literature, dominated the world of rabbinic learning prior to the fall of Jerusalem in 70 AD, is regarded as the Talmudic literature's paradigm of halakhic dispute. Their disagreements extend over three hundred or so matters of halakha. In thirty-four cases, however, apart from stating their different opinions, the texts detail the debate that followed. These allow the reader a fairly detailed glimpse of how the editors and compilers of a broad selection of early rabbinic texts envisaged the two Houses' discursive practices—which turn out to be interestingly different. One such difference is highly significant in the present context. In seven of the fifteen debates in which the Shammaites are accorded the last word in the debate, the Hillelites are described as having abandoned their initial positions, and 'admitting' to those of their adversaries. By contrast, in none of the eighteen cases in which the Hillelites are accorded the last word are the Shammaites ever on record as having changed their mind and adopting the position of their adversaries. Now, a willingness seriously to entertain arguments against one's halakhic legacies, and when convinced to be willing to actually retract and modify them, is the hallmark of antitraditionalism. Constantly aware of their liability, and the system's liability to err, antitraditionalists are constantly on the lookout for possible flaws and problems in what they hold to be the law. Traditionalists, obligated absolutely by their traditions, are, by contrast, unable to do so even when faced with arguments for which they have no answer.

It is clear from this and a variety of other evidence, with which I have dealt elsewhere,[65] that the Talmudic literature pictures the two Houses as archetypal representatives of the two approaches I have been discussing, and, subsequently, as engaged in the very dispute of which we speak. While the Shammaites are depicted as steadfast traditionalists, the Hillelites are portrayed as avid self-critical antitraditionalists, who warmly welcome the constant challenge they receive from their adversaries. As one well-known text puts it, the reason we follow the House of Hillel in matters of halakha is because they were modest, self doubting pluralists—or in the Talmud's own words: "because the were humble and accommodating, and would not only always give the Shammaite position a hearing, but went out of their way to consider it before deciding their own."[66] In the realm of halakha, the antitraditionalist stands before the system of religious law as Moses and Leah are said to have confronted the morality of God's decisions and actions.

A Canon Divided

If there is any truth to my readings of these texts, then there are two foundational theologies, rather than one, hard at work at the heart of the formative canon of the Jewish tradition; each generating a very different notion of religious commitment. One insistently advocating a religious culture of uncritical pious compliance with the religious sources of authority premising faith in a perfect system of law sanctioned by a perfect, intervening God; the other, equally insistently, advocating a religious culture of constant

65 See Shapira and Fisch 1998, and more recently Fisch 2007.

66 Bavli, *Erwin* 13b.
מפני שנוחין ועלובין היו, ושונין דבריהן ודברי בית שמאי. ולא עוד אלא שמקדימין דברי בית שמאי לדבריהן.

normative rethinking that premises systematic denial of the moral perfection of any of the system's sources of authority, God in heaven included. This latter voice, extends to all aspects of Jewish religiosity the open-ended, constructive skepticism that Rabbi Berlin describes in the introduction to his great exegetical work as being analogous to modern science, which remains wholly confined by the first voice (to which Berlin himself subscribes) to biblical exegesis.

Judaism's formative canon is divided, one could say, with regard to its own very meaning and message. And yet, its two contrasting voices somehow remain fully accommodated and contained within a single religious framework. How is that possible, with differences so vast and fundamental? How can such conflicting accounts of religiosity, pertaining to such different theologies, remain authentic expressions of one and the same religious culture? What do they share? What common ground binds them? The answer, I believe, lies in Judaism's unique and undisputed approach to the very notion of a religious canon—the approach expressed so vividly by Rabbi Berlin in comparing the Jewish exegetical tradition to science. Unlike both early modern Christian approaches discussed in the first part of this paper, Judaism makes clear and undisputed distinctions between its holy texts, the religious obligation to study them and make sense of their meaning, and the resulting plethora of readings thereby produced. Its unique feature is that while the texts themselves are solemnly deemed sacred and their study, the very highest of religious works, none of the readings and understandings thereby produced are treated as final or binding. Judaism in its Talmudic constructions succeeds in forming a single and coherent religious entity by becoming a "people of the book"[67]—a people constituted by its books, not by any one of the various readings her books are given. As we have seen, Judaism's foundational project of religious self-understanding produces not one, but two radically different self-readings. The submissive version of Jewish religiosity, we have seen, is an authentic voice of Judaism, whose presence in Talmudic culture is beyond dispute. But by being but one of *two* such voices, it is incapable, in principle, of viably accounting reflectively for the very second-order dispute of religiosity to which it is party. And it is here, with respect to the question of religiosity itself, that the confrontational voice rules supreme, despite its resolute opposition to all manner of final religious ruling.

---

[67] The allusion to Halbertal 1997, is quite intentional. For a friendly critique of the Talmudic chapter of Halbertal's work along the lines of the argument of the present paper, see Fisch 2005.

Unfortunately, rabbinical Judaism has not shown much *religious* interest in science, and no interest at all in understanding its methods or the nature of its success; and needless to say, those who owe their mindset to modern science have shown no interest in rabbinic Judaism and its conceptions of religiosity. The background, cultural context, initial motivations, and dramatic rise of modern science owe exclusively to Christian sensibilities, and its later and cultural impact, to a painful, sobering, keenly critical reassessment of those very motivations and sensibilities. But one cannot help wondering how modernity's culture of science and the various reactions to it might have evolved had that not been the case. Exercises in counterfactual history cannot change what happened, but they can enrich our understanding of why things turned out the way they did. More importantly, such thought experiments offer us glimpses of some of the roads not taken, and of some of the roads we might want yet try. Rabbi Berlin's extraordinary text offers us a fleeting, insufficiently developed, yet suggestive glimpse of a joint approach to science and religion of a kind Western culture has never experienced: viewing them both as constructively skeptical, modestly assertive, and genuinely pluralistic and open-ended undertakings. Grounded in a religiously meaningful, yet humble epistemology, such a joint view of scientific and religious knowledge stands in opposition to both highly worrying forms of latter-day irrationality. In a world caught between uncritical dogmatism, on the one hand, and an equally uncritical radical relativism, on the other, it is a voice much needed both within and without observant Judaism.

Bibliography

1964. Midrash *Devarim Rabbah*, Ed. Saul Lieberman. 2d ed. Jerusalem: Wharmann Books.

1993. *Sifre on Deuteronomy*. Ed. Haim Saul Horovitz and Louis Finkelstein. New York and Jerusalem: The Jewish Theological Seminary of America.

1996. Midrash *Bereshit Rabbah*—Critical Edition with Notes and Commentary, Ed. Jehudah Theodor and Chanoch Albeck. 2d printing, Jerusalem: Shalem Books.

1997. *Mekhilta D'Rabbi Ismael*, Ed. Saul Haim and Israel Abraham Rabin. 2d ed. Jerusalem: Shalem Books.

Bacon, Sir Francis. 1858. Novum Organum. In *The Works of Francis Bacon*, Ed. James Spedding, Robert Leslie Ellis, and Douglas Denon Heath. London: Longman.

——. 2000. *The Advancement of Learning*, Ed. Michael Kieman, Oxford: Oxford University Press.

Berkeley, George. 1734. The Analyst; or a Discourse Addressed to an Infidel Mathematician. In *The Works of George Berkeley, D.D. Bishop of Cloyne*, Ed. George Sampson, London: George Bell and Sons, Vol. III, 1–51.

Berlin, Rabbi Naphtali Zvi, Yehuda. 1986. Commentary on *She'iltot de-Rav Achai Gaon*, 3 Vols. Jerusalem: Mosad Ha-Rav Kook.

Blidstein, Gerald. 1997. 'Even If They Tell You That Right Is Left and Left Right': On Institutional Authority in Halakha and its Limits (Hebrew). In *Between Authority and Autonomy in the Jewish Tradition*. Ed. Zeev Safrai and Avi Sagi. Tel Aviv: Ha-Kibbutz ha-Meuchad, 158–80.
Bono, James J. 1995. *The Word of God and the Languages of Man: Interpreting Nature in Early Modern Science and Medicine. Volume 1—Ficino to Descartes*, Madison: The University of Wisconsin Press.
Brooke, John Hedley. 1991. *Science and Religion: Some Historical Perspectives*. Cambridge: Cambridge University Press.
The Canons and Decrees of the Sacred and Oecumenical Council of Trent. 1848. Trans. J. Waterworth. London: Dolman.
Cohen, L. Jonathan. 1970. *The Implications of Induction*. London: Methuen.
——. 1977. *The Probable and the Provable*. Oxford: Oxford University Press.
——. 1989. *An Introduction to the Philosophy of Induction and Probability*. Oxford: Oxford University Press.
Descartes, Rene. 1989. *Discourse on Method and the Meditations*. Trans. John Vietch. New York: Prometheus Books.
Dugas, René. 1988. *A History of Mechanics*. Trans J.R. Maddox. New York: Dover.
Duhem, Pierre. 1954. *The Aim and Structure of Physical Theory*. Princeton: Princeton University Press.
——. 1980. *The Evolution of Mechanics*. Trans. Michael Cole. Alphen aan den Rijn: Sijthooff and Noordhoff. (first French ed., Paris, 1903)
Feyerabend, Paul. 1977. *Against Method: Outline of an Anarchistic Theory of Knowledge*. London: Verso.
Fisch, Harlod, trans. & ed. 1989. *The Jerusalem Bible. The Holy Scriptures*. Jerusalem: Koren Publishers.
——. 1993. Authority and Interpretation: Leviathan and the 'Covenantal Community.' *Comparative Criticism*, 15: 103–23.
Fisch, Menachem. 1989. The Problematic History of Explaining Science (Hebrew). *Alpayim* 1: 194–208.
——. 1991. *William Whewell Philosopher of Science*. Oxford: Claredon Press.
——. 1997. *Rational Rabbis: Science and Talmudic Culture*. Bloomington: Indiana University Press.
——. 2005. Canon, controverse et réforme; une réflexion sur l'autre voix du judaïsme talmudique. *Les Cahiers du judaisme*, 18: 61–75.
——. 2007. Legislation and Legal Norm Development in Jewish Law. *Periodica de re Canonica* 96: 23–40.
Frazer, James. 1890. *The Golden Bough: The Roots of Religion and Folklore*. 2 vols. New York: Avenel.
Hacohen, Malachi Haim. 2000. *Karl Popper—The Formative Years, 1902–1945*. Cambridge: Cambridge University Press.
Halbertal, Moshe. 1997. *People of the Book: Canon, Meaning, and Authority*. Cambridge MS: Harvard University Press.
Halbertal, Moshe, David Kurtzweil, and Avi Sagi, eds. 2005. *The Discourse of Faith: Essays in Memory of Tal Kurtzweil*. Tel Aviv: Keter.
Harrison, Peter. 1998. *The Bible, Protestantism, and the Rise of Natural Science*. Cambridge: Cambridge University Press.
Herbert, Christopher. 2001. *Victorian Relativity: Radical Thought and Scientific Discovery*. Chicago: The University of Chicago Press.
Howell, Kenneth J. 2002. *God's Two Books: Copernican Cosmology and Biblical Interpretation in Early Modern Science*, South Bend, IN: University of Notre Dame Press.
Jardine, Lisa, and Alan Stewart. 1998. *Hostage to Fortune: The Troubled Life of Francis Bacon*. New York: Hill and Wang.
Lagrange, Joseph Louis. 1788. *Mécanique Analytique*. Paris.

——. 1797. *Théorie des fonctions analytiques,* Paris.

Lindberg, David G., and Ronald L. Numbers, eds. 1986. *God and Nature: Historical Essays on the Encounter between Christianity and Science.* Berkeley: University of California Press.

Lakatos, Imre. 1978. History of Science and its Rational Reconstructions. In *The Methodology of Scientific Research Programs: Philosophical Papers by Imre Lakatos, Vol. I.* Ed. John Worrall and Gregory Currie. Cambridge: Cambridge University Press, 109–38.

Lorberbaum, Menachem. 1995. Learning from Mistakes: Resources of Tolerance in the Jewish Tradition. In *Democratic Education in a Multicultural State* (Special issue of the *Journal of the Philosophy of Education,* vol. 29), Ed. Yael Tamir. Oxford: Blackwell, 115–26.

Menn, Stephen. 1998. *Descartes and Augustine.* Cambridge: Cambridge University Press.

Moore, James R. 1986. Geologists and Interpreters of Genesis in the Nineteenth Century. In *God and Nature: Historical Essays on the Encounter between Christianity and Science,* Ed. David G. Lindberg and Ronald L. Numbers. Berkeley: University of California Press, 322–50.

Newton, Isaac. 1687. *Philosophia naturalis Principia Mathematica.* London.

——. 1704. *Opticks: Or a Treatise on the Reflexions, Refractions, Inflexions and Colours of Light.* London.

Nietzsche, Friedrich. 2001. *The Gay Science: With a Prelude in German Rhymes and an Appendix of Songs.* Cambridge Texts in the History of Philosophy. Ed. Bernard Williams. Trans. Josefine Nauckhoff and Adrian Del Caro. Cambridge: Cambridge University Press.

Popper, Sir Karl Raimond. 1959. *The Logic of Scientific Discovery,* London: Hutchinson. (first German ed., Vienna 1934)

Rorty, Richard. 2007. *Philosophy as Cultural Politics: Philosophical Papers, Vol. 4,* Cambridge: Cambridge University Press.

Safrai, Zeev, and Sagi, Avi. 1997. *Between Authority and Autonomy in the Jewish Tradition* (Hebrew). Tel Aviv: Ha-Kibbutz ha-Meuchad.

Sagi, Avi. 1993. The Canonical Scriptural Text and the Hermeneutical Challenge: A Critical Analysis in the Footsteps of Nahmanides (Hebrew). *Da'at* 50–52: 121–41.

——. 2007. *The Open Canon: On the Meaning of Halakhic Discourse.* London: Continuum.

Sagi, Avi and Zvi Zohar, eds. 2001. *Renewing Jewish Obligation: The World and Thought of David Hartman* (Hebrew). Tel Aviv: Ha-Kibbutz ha-Meuchad.

Sampson, George, ed. 1898. *The Works of George Berkeley, D.D. Bishop of Cloyne.* London: George Bell and Sons.

Schremer, Adiel. 2001. Undermining Interpretation and Interpretive Uprooting (Hebrew). In *Renewing Jewish Obligation: The World and Thought of David Hartman.* Ed. Avi Sagi and Zvi Zohar. Tel Aviv: Ha-Kibbutz ha-Meuchad, 747–69.

Shapin, Steven, and Simon Schaffer. 1985. *Leviathan and the Air-Pump: Hobbes, Boyle, and the Experimental Life.* Princeton: Princeton University Press.

Shapira, Haim, and Menachem Fisch. 1998. The Debates between the Houses of Shammai and Hillel: The Meta-Halakhic Issue (Hebrew). *Iyyunei Mishpat* (*The Tel Aviv University Law Review*) 22.3: 461–97.

Sprat, Thomas. 1667. *History of the Institution, Design, and Progress of the Royal Society of London for the Advancement of Experimental Philosophy.* London.

Tamir, Yael, ed. 1995. *Democratic Education in a Multicultural State* (Special issue of the *Journal of the Philosophy of Education,* vol. 29). Oxford: Blackwell.

Taylor, Charles. 2007. *A Secular Age.* Cambridge, MA: Harvard University Press.

Tirosh-Samuelson, Hava. 1997. Theology of Nature in Sixteenth-Century Italian Jewish Philosophy. *Science in Context* 10.4: 529–70.

Toulmin, Stephen. 1991. *Cosmopolis: The Hidden Agenda of Modernity.* Chicago: The University of Chicago Press.

Urbach, Peter. 1987. *Francis Bacon's Philosophy of Science: An Account and Reappraisal.* La Salles, IL: Open Court.

Whewell, William. 1837. *History of the Inductive Sciences, from the Earliest to the Present Time.* 3 vols. London.

———. 1840. *The Philosophy of the Inductive Sciences, Founded upon Their History*. 2 vols. London.
Worrall, John and Gregory Currie, eds. 1978. *The Methodology of Scientific Research Programs: Philosophical Papers of Imre Lakatos, Vol. I.* Cambridge: Cambridge University Press.
Yadin, Azzan. 2004. *Scripture as Logos: Rabbi Ishmael and the Origins of Midrash*. Philadelphia, PA: University of Pennsylvania Press.

SCIENCE, RELIGION, AND RATIONALITY: A NEO-HEGELIAN APPROACH

Menachem Fisch

Introduction

This paper addresses a major problem with rationality, the solution to which, I believe, has important consequences for the study of both science and religion. It concerns the outermost boundaries of reasoned critique, the point at which one attempts to critically assess the very norms by which such criticism is leveled. Section 1 describes my point of departure with respect to rationality. Sections 2 and 3 present the problem of rational framework transition and my proposed solution respectively. Sections 4 and 5 discuss some of the lessons that adopting such an approach have first for science and then for religion.

Rationality and Criticism

Karl Popper's equation of rationality with criticism is a good place from which to start. Popper's idea is simple and compelling. To act is to introduce some change in one's world or in one's view of it. To refrain from acting is to leave one's world or one's view of it unchanged. To do either rationally[1] is to do so for a reason. To have reason to act is to find one's situation sufficiently lacking to warrant change; to have reason not to act is to deem one's situation preferable to the considered alternatives. Hereby lies the crude rational force of problems and failings that, for Popper, renders problem-seeking and problem-solving definitive of rationality and the cornerstone to his entire philosophy.

* This article was first published in *Toronto Journal of Theology* 29, no. 2 (2013), pp. 319–336. DOI: 10.3138/tjt.1942. Reprinted with permission from University of Toronto Press (www.utpjournals.com).

[1] Underlying my discussion is a firm distinction between viewing rationality as a prospective, evaluative category of act*ing*, rather than as a retrospective evaluative category of *acts* or *moves*, between attributing rationality to exercises of *agency*, as opposed to the quality or worthwhileness of their outcomes.

But his basic identification of rationality and criticism requires further elaboration. First, any rational action thus construed involves some element of *self*-critique. Awakening to a problem necessarily presupposes a measure of both self-reckoning and self-correction. As Popper has taught us, to take reasoned action is in some irreducible sense to acknowledge one's liability to err.

However, a deeper element of critical self-scrutiny is inherently involved in our very capacity to reason. As Christine Korsgaard forcefully argues,[2] to act for a reason requires deciding first what one's reasons are. Not every impulse we have to believe is reason to believe, as she puts it, not every desire is reason to act. Korsgaard thus locates the defining feature of rational agency in the essentially *intra* subjective domain of deliberating which sensations and impulses merit "reflective endorsement" as reasons to believe and to act.

However, thinkers like Wilfrid Sellars and Robert Brandom insist on giving Korsgaard's neo-Kantian picture a sharp neo-Hegelian twist by locating rationality's defining feature in the essentially *inter*subjective discourse of reasons, where players engage in the game of giving and asking for reasons and keep deontic score of each other's commitments and entitlements.

Korsgaard is right to claim that self-constitution belongs wholly and exclusively within the realm of the intra-subjective. No one but oneself can endorse one's reasons to act and be committed to one's norms. No one but oneself can oppose certain of one's desires and impulses and identify with others. Adopting the reflective stance is to stand in critical self-judgment, wary of the persistent claims of our sensations, desires, and impulses to motivate, just as she describes.

Once we have decided on our reasons, we are in a position to rationally exercise our agency in action, which again requires adopting a self-critical stance, this time toward our world and our picture of it. Constituting ourselves as agents and exercising our agency in action are hence profoundly self-critical and self-adjudicating undertakings.

Finally, in engaging the game of giving and asking for reasons, in exposing our first-order Korsgaardian choices to the critical reflection of others, a third and higher form of reflective wariness is added. Taken jointly, the three components of human agency—endorsing reasons, applying them in action, and exposing them to the critical appraisal of others—display our

[2] Christine M. Korsgaard, with G. A. Cohen, R. Geuss, T. Nagel, and B. Williams, *The Sources of Normativity*, ed. O. O'Neill (Cambridge: Cambridge University Press, 1996), 93.

rationality as so thoroughly steeped in self-criticism as to justify viewing the two terms as synonymous, as critical rationalists insist.

Viewing human agency thus also explains why such thinkers as Sellars and Brandom are fully justified in deeming the third component—participation in the intersubjective dialogue of reasons—as the epitome of human rationality, for in subjecting our reasons to the normative appraisal of fellow reasoners, we are *applying our rationality to itself*, as it were, employing others in reflecting critically upon our own attempts at critical reflection, acknowledging our fallibility by engaging their help. This is human rationality at its utmost: when at the very point in which critical reflection comes full circle, it critically engages itself with the help of others. And it is here that, according to Brandom, the "I" becomes "we," as the self awakens to its social and discursive embeddedness. But none of this is simple, for in the vexingly problematic nature of its social embeddedness lies the problem of rationality I wish to address, a problem that involves a crucial aspect of criticism that critical rationalists have tended to overlook.

Normativity and the Problem of Self-Criticism

To find a system or situation wanting and in need of repair is to make a *value* judgment, for one is deeming its present state *inferior* to what one considers a practically achievable alternative. And to commit oneself to putting it right is to direct a similar normative judgment toward oneself by deeming one's present state of inaction inferior to the viable and justified act of intervention one is contemplating. In short, to act for a reason is to take stock of what *is* the case and act in accord with what one believes it *ought* to be, which necessarily presupposes reference to an existing set of relevant values or criteria of propriety. To criticize is to pass normative judgment, for which a normative framework must necessarily already be in place. All criticism is, in other words, inherently framework-dependent in the strong sense of the term—and, yes, one can hear some of the Popperians cringing.[3]

[3] I am not trying to claim that rational action as such necessarily presupposes an organized, coherent, and reflected upon system of values. But I do insist that for action to be deemed even minimally rational, it must be taken for a reason, which in turn requires, by definition, making a value judgment, for which a relevant set of normative commitments must therefore be in place.

At its intra-subjective Korsgaardian level of selecting and endorsing reasons and applying them in action, the normative dependency of rationality is understandable and seemingly unproblematic. Deeming a desire, sensation, or impulse worthy or unworthy of endorsement as a reason, and a state of affairs in need or not in need of correction, are clear cases of norm *application*. And norms thus applied necessarily presuppose the agent's firm prior commitment. Conducting oneself rationally at this level is to hold oneself *accountable to* one's *existing* standards of propriety. But rationality requires of us not only to apply our standards *in* critical appraising but to subject them *to* critical appraisal, not only to act in accord with the standards and norms to which we are committed but to hold those very norms and standards in normative check. But how is that possible? How can the standards to which one is committed be effectively subjected to one's own rational appraisal if it is by means of those very standards that one rationally appraises?

Sellars and Brandom imply that this is precisely what exposing our reasons and actions to others is meant to achieve. But how? If a person is incapable of critically appraising her very standards of appraisal, how could she be convinced to do so by somebody else? Norms may of course be criticized in the light of higher norms—we do it all the time. One may criticize oneself and others for failing to live up to, say, the standards of one's college, and quite consistently criticize those very standards in the light of more fundamental norms, such as those of one's religion, or one's liberal commitments. But there is a deep sense in which cases like these prove only that the lower-level norm, though a convention one holds to, was not really a normative commitment to start with. Normative purging in the light of higher-level commitments is how we find out what we are really committed *to*, not whether what we *are* committed to merits commitment. The standards of propriety to which one is genuinely committed, it would seem, are by definition immune to normative self-criticism. But that seems absurd! John McDowell makes the point forcefully, claiming that, as he puts it, all thinking "is under a standing obligation to reflect about and criticize the standards by which, at any time, it takes itself to be governed."[4] However, no one, including McDowell, has shown how, "from the midst of the way of thinking one is reflecting about," the normative failings of one's normative

[4] John McDowell, *Mind and World with a New Introduction by the Author* (Cambridge, MA: Harvard University Press, 1996), 81.

outlook can in principle be disclosed by subjecting it to its "own reflective scrutiny."

Thinkers like Carnap, Kuhn, Rorty, and Davidson have argued explicitly that doing so is impossible. Others, notably Popper and some of his school, deny there being a problem at all by rejecting the very notion of framework dependency as a "myth."[5] But the majority of relevant writers, Sellars, Brandom, and Korsgaard included, avoid discussing the question by simply not raising it. The notable exception is Michael Friedman's impressive, yet in my opinion ultimately unsuccessful, attempt to account for the rationality of framework transitions in science,[6] to which I shall return. This is no coincidence. The problem of normative self-criticism is real and notoriously difficult.

In the Company of Critics

And yet I believe it can be solved, though not by the kind of intra-subjective self-distancing described by McDowell. In that case, the key must clearly lie in exposure to the normative critique of others, as Michael Walzer's work on social criticism,[7] and that of Sellars and Brandom briefly mentioned, clearly imply. But we need to tread carefully. For a framework transition to be rational, it must first be normatively faulted by those whose framework it is *while still in play*, and only then be abandoned for an alternative they consider to be better. Wittgensteinian "persuasion,"[8] Rortian irony,[9] and Kuhnian conversions or gestalt switches[10] won't do, because they reverse the process: the framework is first replaced, and then the transition is rationalized retrospectively.

---

[5] See Karl R. Popper, *The Myth of the Framework: In Defense of Science and Rationality* (London: Routledge, 1994).

[6] See especially Michael Friedman, *Dynamics of Reason: The 1999 Kant Lectures at Stanford University* (Stanford: CSLI Publications, 2001); and Friedman, "A Post-Kuhnian Approach to the History and Philosophy of Science," *Monist* 93 (2010), 497–517.

[7] Michael Walzer, *Interpretation and Social Criticism* (Cambridge, MA: Harvard University Press, 1887); and Walzer, *In the Company of Critics: Social Criticism and Political Commitment in the Twentieth Century* (New York: Basic Books, 1988).

[8] Ludwig Wittgenstein, *On Certainty*, ed. G. E. M. Anscombe and G. H. von Wright, trans. Denis Paul and G. E. M. Anscombe (repr.; Oxford: Blackwell, 1974), 81e, ss. 611–12.

[9] Richard Rorty, *Contingency, Irony, and Solidarity* (Cambridge: Cambridge University Press, 1987).

[10] Thomas S. Kuhn, *The Structure of Scientific Revolutions*, vol. 2, no. 2, *International Encyclopedia of Unified Science*, 2nd ed. (Chicago: University of Chicago Press, 1970), esp. 94, 150–51.

Neither are attempts to show that a functioning framework can be normatively faulted by comparing it to other frameworks seemingly deemed to be normatively superior. For reasons I cannot enter into here, this type of scenario is philosophically incoherent.[11]

What is needed, then, is a way of explaining, not how one can be won over by an alternative form of life, but how one can be *motivated* to do so by criticizing the set of values to which one is still committed. But this, we have seen, is, in principle, humanly impossible, for how can a constitutive framework be impeached by the very principles of rational reasoning and reckoning of which it is constitutive? Nor can we be motivated to do so by normative criticism leveled at us by others—or at least so it seems, for how can someone else's argument transcend the seemingly impenetrable glass ceiling of one's own self-reflection?[12] And yet I believe it can, but to see how, we need to take a closer look at prudent criticism and how it works.

Unlike doubt, criticism aims not merely at raising the possibility but at *proving* that something is sufficiently amiss to require its addressees' attention. Criticism hence aims not only at being heard and understood but at being *endorsed* by its addressees as *self*-criticism and must, therefore, be leveled not from the critic's but from her addressees' perspective, emulating, as best it can, *their* world view and values, mounting its case on premises that *they* hold true and in ways *they* deem valid.

And the same applies to criticizing someone else's norms. The problem, as we have already seen, is that it is impossible, in principle, to frame such a critical argument from that person's perspective, for there is never available a set of premises he is liable to consider true, that can be shown, convincingly *for him*, to entail a denunciation of his very norms.[13] But if it is

---

[11] That is because sufficiently diverse normative outlooks will never share the set of thin evaluative concepts such as "good," "inappropriate," and "wanting" required for such comparisons. See Menachem Fisch and Yitzhak Benbaji, *The View from Within: Normativity and the Limits of Self-Criticism* (South Bend, IN: University of Notre Dame Press, 2010), especially chaps. 1–3.

[12] No problem arises with regard to the rationality of *supplementing* one's normative outlook by previously unconsidered possibilities—like a non-Muslim being struck by traditional Islam's norms of hospitality. The problem I am addressing is that of rationally *replacing* or modifying *existing* norms, which, as we have seen, requires adopting a normatively critical stance toward one's normative commitments. This, I insist, is impossible.

[13] Again, this is not an argument against the possibility of self-criticism or endorsing criticism leveled by others in general. Critiques of our beliefs, techniques, methods, habits, and actions can be extremely effective. The problem to which I point arises with respect to the normative criticism of norms. My argument thus premises a categorical distinction, wholly ignored by Popper, between the rational acceptance and dislodging of belief, and

impossible in principle to formulate a normative critical argument capable of convincing him, how can such an argument ever be said to have the desired rational transformative effect of moving him to rethink his norms!

What needs to be realized is that prudent normative critics are to some extent aware of much of this. They know (or at least instinctively sense) that levelling their arguments squarely from within their addressee's worldview is impossible. But they also know (or at least sense) that to make its mark, their criticism must be leveled, as far as possible, on the basis of premises their addressee can recognize as his own. What normative critics normally do, therefore—and this is the crucial point—is frame their arguments to some extent *untruthfully*, arguing from a perspective very close to that of their addressee, yet sufficiently (and subtly) different from it to be able to at least mount a valid argument. Arguing from the left, for example, critics will surreptitiously premise certain liberal norms to make their case, while those arguing from the right will tend to smuggle in just enough conservative value to make their arguments stick.

In other words, honest normative criticism challenges certain of its addressee's norms by premising a part-portrayal of her normative framework, which, though largely true, differs noticeably from her own self-image. Since leveled against heartfelt norms, the argument itself, as we have seen, will inevitably be dismissed. And yet the picture of its addressee's normative identity conveyed by the argument's premises may linger on and register, uncoupled from the argument itself.

Such deontic portrayals need not be deemed true for them to attract our attention. In order for them to resonate and disturb, it is enough that we consider our critics *sincere* as we awaken to the realization that they see us differently from the way we see ourselves. Since the points of divergence between our own normative self-portrayal and that conveyed by our critic's argument will pertain to the very norms her criticism calls into question, their incongruity, not unlike as in the case of a disturbing playback device, may well destabilize our initial commitment and render us *ambivalent* toward those norms. And norms to which one becomes ambivalent, as Harry Frankfurt insists, lose their wholehearted volitional grounding, and with it their status as firm commitments.[14] Once demoted, they will

---

the rational "adoption" and dislodging of *normative commitment*. Internalizing a norm and rendering it one's own is a process very different from being convinced of the truth of a statement or the usefulness of a tool.

[14] See Harry G. Frankfurt, *The Reasons of Love* (Princeton, NJ: Princeton University Press, 2004), 91.

be subjected, as a matter of course, to the normative critical scrutiny of the self's surviving commitments. It is in this way, I submit, that exposure to the echo chamber of trusted normative criticism *is* capable of inducing the levels of inner discordance and self-alienation necessary for truly transformative normative *self*-criticism.

Paradoxically, the point at which rationality asserts itself most profoundly is thus precisely the point, pace Sellars and Brandom, at which the rational, *inter*subjective game of levelling and responding to criticism exhausts itself, for it cannot be a point of discursive continuity. The neo-Kantian realm of normative self-criticism and self-governance is *not* a continuation or simple internalization of its external, neo-Hegelian counterpart. It is prompted by it, *caused* by it, if you wish. And yet, I insist, it is rational, at least as rational as the potentially ambivalating, self-correcting, indispensable effect of a playback device.

So much, if very briefly, for the problem I see with rationality and for the neo-Hegelian solution I have proposed. Now to their significance, first to science and then to religion.

Lessons for Science

Faced by the revolutionary history of each of the modern natural and human sciences, philosophers of science could not as easily evade the question of the rationality of framework transitions as do the vast majority of philosophers of language, mind, and self—certainly not after Kuhn's *Structure of Scientific Revolutions*. Kuhn's famous work offered a vivid community-, language-, and practice-based, neo-Kantian portrayal of the firm framework dependency of scientific knowledge and research, that could not be ignored—a picture that, as Kuhn's most prominent latter-day follower and critic, Michael Friedman, puts it, in its inability to account for the rationality of scientific paradigm shifts, fiercely challenged "the ultimate rationality of the scientific enterprise." If in paradigm shifts the very criteria of scientific propriety are replaced, there seems "no sense left in which such transitions can still be viewed ... as based on good [scientific] reasons."[15]

[15] Friedman, *Dynamics of Reason*, 47. Neo-Kantian analyses of framework transitions in science, both philosophical and historical, ranging from Reichenbach and Schlick's studies of relativity, through the later Carnap's theory of linguistic frameworks, and Kuhn's work on scientific revolutions, to Friedman's dynamics of reason render such attempts to deny the "subjectivist," framework dependency of human judgment at its best, as in Nagel's *The Last Word* (Oxford: Oxford University Press, 2001), largely untenable.

Friedman's alternative solution, to which I cannot enter here in any detail, is also highly problematic and has been forcefully contested.[16] He himself seems also to have had second thoughts and in his extensive and most recent work takes back much of his original account.[17] But if there is any truth to the solution I propose to the problem of normative self-criticism, a different, less problematic, and more broadly applicable account of scientific paradigm shifts is forthcoming.

It describes such transitions as two-stage processes, in which practitioners of voice and standing are first "ambivalated" by critics, and later (inadvertently) promulgate their newfound indecision within their communities by means of a particular form of published work. Both stages represent irreducible moments of intense *intra*-subjective deliberation that do not admit easily to the collectivist modes of explanation that have come to dominate science studies since Kuhn.

Truly transformative normative self-criticism is possible, we have seen, but it depends necessarily on an environment of potentially "ambivalating" trusted criticism. But who are a science's trusted critics? With all due respect to Walzer's insistence on the "connectedness" of effective normative criticism,[18] given science's firm framework dependency, its trusted normative critics must come from the outside.

Peter Galison's novel notion of the scientific trading zone is useful in this respect.[19] Scientific trading zones are the "locales" of professional engagement outside one's home community, where one bids for financial support, "trades" with neighbour disciplines for techniques or instruments, offers scientific opinion, engages students, the media, the public, or the historians, sociologists, and philosophers of one's field. Galison nicely portrays the professional, interest-driven purposefulness of such interactions and how

---

[16] See especially Fisch and Benbaji, *View from Within*, chap. 5; Menachem Fisch, "Toward a History and Philosophy of Scientific Agency," *Monist*, 93 (2010), 518–44; Don Howard, "'Let me briefly indicate why I do not find this standpoint natural,' Einstein, General Relativity, and the Contingent A Priori"; and Robert DiSalle, "Synthesis, and the Synthetic A Priori, and the Origins of Modern Space-Time Theory," both in *Discourse on a New Method: Reinvigorating the Marriage of History and Philosophy of Science*, ed. Mary Domski and Michael Dickson (Chicago: Open Court, 2010), 333–55 and 523–51 respectively.

[17] Michael Friedman, "Synthetic History Reconsidered" in Domski and Dickson, ibid., 571–813.

[18] Walzer, *Interpretation and Social Criticism.*

[19] Peter Galison, *Image and Logic: A Material Culture of Microphysics* (Chicago: University of Chicago Press, 1997), chap. 9. The account of rational framework transitions in science outlined here is argued for and exemplified in some detail in my *Creatively Undecided: Toward a History and Philosophy of Scientific Agency*, forthcoming from the University of Chicago Press.

disciplines are enriched by the findings of other disciplines. But he nowhere asks if and how practitioners might find themselves *challenged* in such settings. He portrays them, like real trading zones, as providing "safe space[s] of shared phenomenology," cordoned off from the parties' differing core commitments by the scientific pidgins and creoles they employ. But I think that he takes his metaphor a little too literally and misses something important.

The highly relevant aspect of "talking shop" to the uninitiated that Galison misses is how, when abroad, practitioners are frequently required to articulate elements of their worldview that are so taken for granted back home that they remain there largely unarticulated. When presenting to others, one is often required to explicate the basics of one's field and is frequently exposed to the friendly and trusted questioning of genuinely curious professionals on whose thinking the framework constitutive of one's own field may have a much weaker grip. Here, trading and conversing outside their field, is where practitioners are liable to experience the destabilizing and "ambivalating" force of external criticism.

But to set a framework transition in motion, their ambivalence and uncertainty needs to be effectively disseminated among the members of their home community. The problem is that hesitance and uncertainty are not normally considered the stuff of scientific publication and are usually relegated to the personal diary and private correspondence, which do not carry their master's voice very far. Yet at times the doubts and indecision of such individuals come tacitly to inform their creative efforts to overcome them.

Such efforts, I submit, typically take the form of ambitious, yet uneasily split hybrid attempts to re-represent their field's basic assumptions that combine elements of the old framework with an imaginative groping toward new possibilities. When analyzed *prospectively*, such works can be seen to represent anxious, unstable, yet highly creative departures from heartfelt commitments capable of motivating others to seek cleaner, more radical breaks with the old. Classical examples are Tycho Brahe's part geocentric, part heliocentric planetary theory, the combination of Aristotelian and anti-Aristotelian components of Galileo's analysis of projectile motion, and Poincare's Kantian yet non-Kantian geometrical conventionalism. Less known, but equally vivid examples are some of the early Victorian British cases I have worked on in the past, to which I am returning in my current work: George Peacock on algebra, John Herschel and W. R. Hamilton on mathematical physics, and the unholy joining of Kantian and empiricist elements in William Whewell's philosophy of science, whose hybrid influential works unwittingly preserved, and therefore propagated, the keen ambivalence that begot them, thus prompting others to take a clearer and firmer stand.[20]

Lessons for Revealed Religion

Truly religious people are deeply concerned with doing what they believe is religiously right. They constantly ask themselves two kinds of self-searching questions, while refraining almost entirely from asking a third. Most of the time they worry about living up to what they *know* is right. The obligatory status of confession in Catholicism, *tawbah* in Islam, and *teshuvah* in Judaism firmly attests to the permanently felt tension, so prominent in religious life, between one's image of one's ideal religious self and the not always pretty reality of the self as one finds it—religion's analogue to Harry Frankfurt's distinction between first- and second-order volitions.[21] Religious people confess and repent when they knowingly fall short of what they know they should have done or have refrained from doing.

But at times, religious people direct their self-questioning more profoundly toward their ideal rather than empirical selves. We are doing our best to do what we *think* is right, religious people often ask themselves, but are we right in thinking so? Have we got what's right *right*? Such higher-level concerns represent a deeper form of religious self-reckoning, and though they rest on less firm ground than the first kind of self-questioning, they represent a more serious commitment to religious truth. Such questions, of course, are far easier to ask than to answer, for to doubt what one truly believes is the right thing to do is to question one's very religious commitments, one's very norms of religious propriety. It is extraordinarily difficult, as we have seen, to create real distance between the sacred text and our reading of it, because it is our very understanding of it that manifests its sacrality for us. To take an obvious example, Jews and Christians read certain of the Old Testament's halakhic passages very differently. Each can fully understand how the other reads each verse, but neither is able to rationally assess whose interpretation is better, for the simple reason that each

20 My forthcoming *Creatively Undecided: Toward a History and Philosophy of Scientific Agency*, returns to the first of these examples in some detail.

21 The idea, first proposed in his "Freedom of the Will and the Concept of a Person," *Journal of Philosophy* 67 (1971): 5–20, serves as a grounding of his work on agency since.

uses the thin evaluative term *better* as shorthand for the very interpretive standards that yield their different readings in the first place.[22]

Cast in the terminology developed earlier in this paper, the constitutive internalized set of religious norms and values that furnishes a person's reasons to believe, understand, and act do not reside in God's Word in itself (the Kantian turn of phrase is deliberate, of course), but in what that person prudently takes his Word to be. She may doubt that she has got it right, but, as we have seen, she is powerless to criticize its religious worth. However, here too, as I have argued, exposure to the trusted, potentially ambivalating critique of members of other communities may enable her to transcend the restrictive boundaries of her normative self-questioning. Hence the crucial *religious* importance of critical interreligious discourse, which, for the reasons explained above, is the only viable way for the religiously committed to hold their systems of religious value in normative check. Since my argument thus far has been quite general, what was said above about the normatively self-enriching potential of the scientific trading zone applies just as well to the importance of what might be termed the religious trading zones of interreligious dialogue.[23]

But it is the third, virtually no longer existing type of religious self-questioning that interests me most. It pertains to the vertical, rather than to the horizontal axis of religious discourse, representing a concern not with the extent to which one *has got* the Good Word right, but with the extent to which the Good Word *is* right! When wary of the quality of their *understanding* of God's Word, true adherents of revealed religion take the moral perfection of both the Word and its author for granted. God's infallible moral flawlessness is not merely assumed or taken as true *in* the religious discourses of God-centred communities, but functions as a synthetic a priori framework proposition constitutive *of* all such latter-day discourse. God is not *found* to be all-good but is *presupposed* necessarily to be so. The question of God's moral perfection is not raised, therefore, because raising it subverts the very idea of religiosity, nay, the very idea of God, if you wish. If raised by external critics, it is deemed an attack on religion, while raising it *within* the religious framework borders not on the false or the aggressive, but on the incoherent.

---

22 See note 11.

23 In which I, of course, include dialogue with the non-religious. See note 33 and the text to which it is appended.

Judaism's formative canon, to use Moshe Halbertal's term,[24] tells a very different story, however, for the rabbinic literature of late antiquity, that of the Talmudic period, is sorely divided with regard to the very question of the moral perfection, not only of God, but of all of Judaism's sources of religious authority, and is hence sorely divided on the very notion of religiosity itself. My aim in the following pages is to acquaint ourselves a little with this long-lost position, mainly by means of two potent examples, and to briefly ponder its significance.

What renders the point a little difficult to introduce is that, as mentioned, the master norm of religious disposition is now taken by all forms of God-centred religiosity in all three Abrahamic faith traditions to be that of pious submission—certainly with respect to the divine, for what is it to be genuinely religious, it is argued, if not to submit oneself to God's will by committing oneself to honestly discerning what one's religion requires of one, and undertaking to prudently and piously comply? Needless to say, disagreement abounds as to *what* exactly one's religion requires, but such disagreement, so it is argued, cannot extend to the acknowledged sources of the religion's authority themselves. Frequently debated is the precise meaning of an acknowledged authority's ruling or teaching, but never the ruling or teaching themselves.

Indeed, a major voice of the Talmudic literature defines religiosity in exactly these terms. However, and this is the main point in what follows, it is not the only voice of that literature. Submissive religiosity is rigorously and systematically contested in all its forms and on all levels—firmly, explicitly, and self-consciously. And it is contested on the basis of an explicit denial of the theological meta-principle that grounds it, namely, that of divine perfection, especially divine moral perfection!

The master norm of religious disposition advocated by this voice is, by contrast, that of critical engagement. The stance believers are required piously to adopt toward all four main sources of religious authority—Scripture, the halakic tradition, the halakic authorities, and the Almighty himself—is *confrontational* rather than submissive. One's religious obligation is to constantly hold the rulings and teachings of all four in review and to question and dispute them whenever one deems them to be religiously inappropriate or morally wrong. It is not the four sources' *authority* to rule and to teach that is questioned, but the religious appropriateness

---

24 Moshe Halbertal, *People of the Book: Canon, Meaning and Authority* (Cambridge, MA: Harvard University Press, 1997), 3–6.

and moral worth of their rulings and teachings. What makes this voice fascinating from a theological point of view is that there is nothing religiously *subversive* about it. The Talmudic literature's confrontational theology (not unlike that of Job) is advocated as a form of religious faithfulness and obligation. One is urged to contest and question one's religion's teachings as a matter of deep and serious religious *duty*!

My first example concerns two very different Midrashic renditions of the same rather surprising biblical moment. In the course of his long farewell speech recorded in Deuteronomy, Moses recalls how upon nearing the Holy Land he was ordered by God to "begin to possess" Sihon the Amorite, king of Heshbon, and his land, "and provoke war with him" in order "to put the dread of you and the fear of you upon the nations that are under the whole heaven." Moses apparently chose to disobey, deciding, according to the very next verse, to send "messengers out of the wilderness of Kedemoth to Sihon king of Heshbon with words of peace, saying: 'Let me pass through your land; I will go along by the high way, I will neither turn to the right hand nor to the left.'"[25] The Midrashic compilation *Deuteronomy Rabba*, explains Moses's apparent insubordination:

> Although the Holy One, blessed be … told him "Make war," he sought peace! [He did so, however,] because it is written in the Torah: "When thou comest near to a city to fight against it, then proclaim peace unto it" [Deut 20:10], that is why he sent words of peace to Sihon, for it is written "out of the wilderness of Kedemoth," on account of words of the Torah that come before those of the Almighty. Therefore it says "words of peace."[26]

This is as submissive a reading the verses can be given without doing them real violence. Moses is described as apologizing piously for not heeding the Lord's instruction to provoke war, arguing that he is not at liberty to defy the Torah's explicit commandment always to propose peace prior to attacking. In case of contradiction, he explains, the words of the Torah are to "come before" those of God. Even within a thoroughly submissive religious world view, questions will always arise regarding possible or actual conflicts between its various sources of authority. According to this rabbinic

---

[25] Deuteronomy 2:24–28. For a fuller treatment of the two examples, see the latter part of the previous essay.

[26] *Devarim Rabbah*, ed. Saul Lieberman (Jerusalem: Wahrmann, 1964), 29–30. The last sentence of the paragraph clearly appears corrupted in its present form: "From words that came before the Torah, from words of God. Therefore it says 'words of peace.'" It should be corrected as proposed as a play on the word *kedemoth* read as a derivative of the Hebrew verb *k.d.m.*—precedes, comes before.

rendering, the question Moses faced and answered was to which of the conflicting sources of authority should he properly submit: the Torah that instructs always to first seek peace or the Almighty who instructed to provoke war? Moses opts for the former but morally questions neither.

Numbers Rabba to Numbers 19:20 takes a very different approach, describing the incident incredibly as follows:

> This is one of three things said by Moses to the Holy One, blessed be He, to which the latter replied: "You have taught Me something!"... when the Holy One, blessed be He, said to him: "Make war with Sihon. Even though he does not seek to interfere with you, you must provoke war with him"... Moses did not do so, but, as it is written further down, "sent messengers [... with words of peace]." The Holy One, blessed be He, said to him "By your life! You have spoken well! You have taught Me something! And I shall thereby cancel My words and adopt yours"; hence it is written: "When thou comest near to a city to fight against it, then proclaim peace unto it" [Deut 20:10].[27]

Unlike the submissive Moses of *Deuteronomy Rabba*, Moses is here forcefully portrayed as challenging God's decree by refusing to provoke war with the Emorites on moral grounds. God, in response, is said not to have demanded obedience but to have heeded the objection of Moses and to have introduced the required changes in the written Torah! According to the first Midrash, Moses disobeyed God's explicit command because the Torah instructs differently; according to the second, the Torah ended up instructing differently because Moses disobeyed!

The other two examples cited by the Midrash are equally dramatic. According to one of them, upon hearing the Almighty declare at Sinai in the Second Commandment "for I the Lord thy God am a jealous God, punishing the iniquity of the fathers upon the children unto the third and fourth generation," Moses firmly objected: "Sovereign of the universe! Many are the wicked who have begotten righteous sons. Shall the latter bear some of the iniquities of their fathers?... Is it proper that the righteous should be punished for the iniquity of their fathers?" To which God is reported to have replied as above: "You have taught Me something! By your life! I shall cancel My words and adopt yours!,"[28] for indeed it was (thereafter) written, "Fathers shall not be put to death for children, neither shall children be put to death for fathers, every man shall be put to death for his own sin"

---

27 The Amoraic Midrash to Numbers, *Bamidbar Rabba* (Vilna), 19:33.

28 Ibid.

(Deut 24:16). As this Midrash has it, the principle of individual liability was introduced into the Torah by God as a result of Moses's forceful criticism of the Ten Commandments! Here again God is astonishingly described as authoring a morally deficient early draft of the Ten Commandments, Moses is portrayed as having refused to comply, and having challenged it on moral grounds, and the written Torah, the very Word of God, is subsequently described as having been happily revised by virtue of God's acknowledging and endorsing Moses's moral criticism! The theology and subsequent notion of religiosity at work here differ radically from the kind of pious submission to the High one normally associates with revealed religion. The Midrash's explicit denial of God's moral perfection, and consequently that of the Torah, is by no means regarded an attack on or a subversion of religion. The God described is a covenanting God, but the covenant his human partners are invited to enter is not one of religious obedience and surrender, but one better described as a covenant of constructive confrontation, especially moral confrontation—a sacred partnership dedicated to the betterment of the world, the betterment of the Torah, and the betterment of the self. God's need for the critical input of his human partners is portrayed as being as real as theirs should be for his. God makes known his imperfection by initiating real engagement, by learning, by admitting his mistakes, and recording the corrective process as part and parcel of the written Torah!

In my second example, the voice of religious submission again speaks loud and clear, and again, does not speak alone. Self-consciously opposing it, one finds clear expression of an avid confrontational approach even to hakakha. The dispute in question concerns the religious appropriateness of prayers of anguish. The first text, taken from chapter 9 of Mishna, *Berakot*, contains one of the keenest expressions of submissive religiosity to be found in the entire Talmudic literature. The chapter is devoted to obligatory expressions of awe, wonder, and gratitude in acknowledgement of God's awe-inspiring presence and good work in the world. It commences with the obvious blessings one is supposed to utter when visiting places "where miracles have been wrought for Israel" or "from which idolatry has been extirpated," or when witnessing the likes of shooting stars, earthquakes, and thunderclaps, or on seeing great mountains, hills, seas, and rivers. But the Mishna's submissiveness becomes cruelly apparent when it moves from the wonderful and the awesome to the catastrophic. "For rain and for good tidings one says: 'Blessed [art thou, O Lord our God, King of the universe] who is good and bestows good.' For bad tidings one says: 'Blessed [art thou...] who is the true judge'... Over the dreadful a blessing is said similar to that over the good, and over the good, a blessing is said similar to that over dreadful."[29]

Standing before God, the Mishna rules, the appropriate religious response to suffering is one of capitulation, of total acknowledgement of his supreme judgment. Pain and suffering exist in the world, it implies, but there exists no evil, mischief, or oversight. People may scheme to do evil, natural catastrophes may hit, accidents will happen, but their outcomes, dreadful and unfair as they may seem, are forever the doings of "the true judge," the perfect workings of divine justice. Standing before God in response to tragedy, one should, therefore, not merely resign oneself to one's fate, but be thankful for it. Always. Consequently, the Mishna declares petitional prayers of anguish to be religiously inappropriate if uttered retrospectively in tortured response to past occurrences—not because God is powerless to intervene, but because in a world overseen by a perfectly just, personally intervening deity, all that occurs is deemed to be just by the very fact of its occurrence! Prayers of appeal are appropriate only prospectively, instructs the Mishna, in fearful anticipation of what the future might hold, never after the event.

> For to cry out (in prayer) over the past is to utter a prayer in vain. For instance, If a man's wife is with child and he prays: "May it be Thy will [O God] to grant that my wife bears a male child" this is a prayer in vain. [Similarly] if a person is coming home from a journey and he hears cries of distress in the city, and he prays: "May it be Thy will [O God] to grant that this is not in my household," this is a prayer in vain. For that which has past, one should at all times give thanks, and cry out [only] for that which is yet to come.[30]

In its commentary to the story of the birth of Leah's daughter, Dinah in Genesis 30, the *Tanhuma*, the late fifth- or early-sixth-century Palestinian Midrashic compilation, confronts both the Mishna's ruling and implied theology head-on. The *Tanhuma* openly questions the Mishna's ruling on the religious appropriateness of prayers of anguish.

> "And afterwards she [Leah] bore a daughter" [Genesis 30:21]. May our Rabbi teach us: If a man's wife is with child is one allowed to pray "May it be Thy will that my wife bears a male child"—for so teaches [the Mishna]: "If a man's wife is with child and he prays: 'May it be Thy will [O God] to grant that my wife bears a male child' this is a prayer in vain"? R. Huna in the name of R. Yossi says, despite the fact that [the Mishna] teaches that "If a

29 Mishna, *Berakot* 9:1–2.

30 Ibid., 9:3–5.

> man's wife is with child and he prays: 'May it be Thy will [O God] to grant that my wife bears a male child' this is a prayer in vain," **[the law] is not so**. Rather, one may pray for a son even as she is giving birth, because it is not beyond God's power to change females to males and males to females.

The *Tanhuma* is not suggesting that the Mishna's ruling on the matter was misunderstood or misquoted, but boldly asserts that it is wrong! And it does so not by piously appealing to a higher halakic authority, but by openly contesting the Mishna's theological reasoning. The question, the *Tanhuma* dramatically implies, is not a technical one, but pertains to the very foundation of the religion. "For so says Jeremiah," it continues:

> "Then I went down to the potter's house, and, behold, he was at work on the wheel. And the vessel that he was making of clay was spoiled in the hand of the potter: so he made it again into another vessel, as it seemed good to the potter to make it. Then the word of the Lord came to me, saying, O house of Israel, cannot I do with you as this potter? said the Lord. Behold, as the clay in the potter's hand, so are you in My hand, O house of Israel" [Jeremiah 18:3-6]. Says the Almighty to Jeremiah: if a potter can break a cruse after he has made it and make it into something else, can I not do likewise to you O House of Israel!?

The *Tanhuma* reads the Jeremiah verses as claiming, not that God can reverse at will everything he does, but that not everything He does comes out good, resembling an imperfect potter whose handiwork is not always up to par. Moreover, the *Tanhuma* clearly implies that in addition to his handiwork being occasionally flawed, unlike the typical potter, God needs the flaws to be pointed out to Him. This, as the subsequent passage makes clear, is the theology that lies at the heart of the *Tanhuma*'s radically different approach to petitional prayer, and subsequently, to the religious covenantal relationship in general.

> And similarly in the case of Leah, who after bearing [Jacob] six sons, saw prophetically that twelve tribes were to emerge from Jacob. Having born six already, and pregnant for the seventh time, and with two sons already born to each of the two maidservants, ten had already been born. Therefore, Leah stood, **angrily confronting** the Almighty, saying: Lord of the Universe,... If this [unborn] child [I am carrying] is also a son, then my sister Rachel's share will not [even] be that of the maidservants!

At this point the Midrash could have easily described God as scolding Leah in the way he had scolded the laughing Sarah who had doubted his promise of a child (Genesis 18:14). "Do I not know that this would unfair to Rachel?! Of course you are carrying a Girl!" But the *Tanhuma* opts for the far more radical option. God is depicted as having erred in indeed giving Leah a seventh male child. Realizing his mistake, we are told, "The Almighty immediately responded to her prayer and the boy in her womb was rendered a girl ... And why did she name her (daughter) Dinah? Because Leah the righteous called the Almighty to task."[31]

Here again, God is astonishingly described as being corrected on moral grounds, admitting his mistake, and righting the morally flawed situation for which he was responsible. But unlike my first example, the *Tanhuma's* main lesson is halakhic. Unlike Moses's criticisms of God's rulings of my first example, Leah's confrontation related here is cited explicitly as a model of religious conduct. According to the *Tanhuma*, one is not supposed to suppress prayers of anguish or to always accept the dreadful as the work of the perfect judge. On the contrary, God, the *Tanhuma* all but states openly on the basis of the Leah story, is simply *not* the all-perfect judge and should not be treated as such by anyone!

Conclusion

In other forms of normative discourse—ethical, political, scientific—the neo-Hegelian, intersubjective corrective that enables us to transcend the normative limits of self-criticism, and hence of rationality, requires extending our exposure to the critical scrutiny of others *horizontally*, beyond the consensus-governed boundaries of our home communities. Rabbi Jonathan Sacks writes passionately about the dignity of difference.[32] I would like to emphasize not so much the dignity as the unique *value* of normative otherness, for only in the company of truly external critics can the potentially ambivalating process of serious normative rethinking begin. The rabbinic literature of late antiquity was well aware of this too, locating the supreme moral test of our halakhic presuppositions in how they are perceived *from the outside*, by prudent gentiles.[33] And similar to the voice we have now encountered, neither does that voice stand alone. The two positions belong on the same side of the great Talmudic divide in both univocally denying the system's moral perfection and moral self-sufficiency even at its most basic levels, and in locating the only possible path to improvement in keen

---

[31] Midrash *Tanhuma, Vayetze* 8. The Hebrew subtly makes use of two wordplays see above p. 105, n. 62.

[32] Jonathan Sacks, *The Dignity of Difference: How to Avoid the Clash of Civilizations* (London: Continuum, 2003).

[33] See Menachem Fisch, "Legislation and Legal Norm Development in Jewish Law," *Periodica de re Canonica* 96 (2007), 23–40.

intersubjective normative critique. However, the passages we have studied here are unique in the way they press our moral disagreements with God, rather than those we have with other people, in the service of *mutual* religious self-amelioration—so much so that the normative criticism across the vertical revelatory divide proves even more radical and more effective than its horizontal counterparts—and not because God is assumed to know better. Quite the contrary!

These moral confrontations with a morally imperfect God—and there are many more where my two examples come from!—beautifully and most strikingly bear witness to the way Hegel's original imagery of normative dispute as an arm-wrestling power struggle for domination between masters and slaves to be, morphs already in the Talmudic literature into the kind of a rational dynamic of reasoned discourse one is beginning to encounter in recent rereadings of Hegel.[34]

[34] See, for example, Robert P. Pippin, *Hegel on Self-Consciousness: Desire and Death in the Phenomenology of Spirit* (Princeton: Princeton University Press, 2010).

JUDAISM AND THE RELIGIOUS VALUE OF DIVERSITY AND DIALOGUE: DRAFTING A JEWISH RESPONSE TO *NOSTRA AETATE*

Menachem Fisch

The Challenge of Vatican II

The Decrees, Constitutions and Declarations comprising The 2nd Vatican Council of the mid-1960's combine to constitute a major reform of Catholic belief, practice and governance. But for a religiously engaged, non-Catholic like myself, two of the documents stand out as particularly striking: the "Declaration on Human Freedom" entitled *Dignitatis Humanae*, and the "Declaration on the Relation of the Church to Non-Christian Religions" entitled *Nostra Aetate*.[1]

Dignitatis Humanae is a passionate affirmation of the right of all people to religious freedom, declaring the protection and respect of that freedom a solemn religious duty. "Injury ... is done to the human person and to the very order established by God ... if the free exercise of religion is denied in society," it states. All religions are emphatic about demanding religious liberties for *themselves*, but the central plea of this remarkable document is not self-serving but *self-directed*. Because "this doctrine of freedom has roots in divine revelation," it instructs Christians "at all times to refrain from any manner of action which might seem to carry a hint of coercion or of a kind of persuasion that would be dishonorable or unworthy.... Such a manner of action" it deems to be "an abuse of one's right and a violation of the right of others."

It is extremely difficult to square one's absolute commitment to the truths of one's own faith, with a genuinely religiously sanctioned obligation to respect and protect the right of others to live by their very different, and from one's own perspective, gravely mistaken beliefs. Nonetheless, *Dignitatis Humanae* resolutely affirms the right to freedom of religion,

---

* This chapter will be published as: Menachem Fisch, "Judaism and the Religious Value of Diversity and Dialogue," in *Diversität – Differenz – Dialogizität. Religion in Pluralen Kontexten*, eds. Stefan Alkier, Michael Schneider, and Christian Wiese (Berlin and Boston: de Gruyter, forthcoming), and is printed here with permission from de Gruyter.

[1] All citations from the two documents are from the official vatican translations at: http://www.vatican.va/archive/hist_councils/ii_vatican_council/.

not as a matter of convenience, or even of concern for the sake of peace, but as matter of *religious value.* However, *Nostra Aetate* goes an extraordinary step further. To see how, it is important to realize that an obligation, however serious and deep, to grant people the *right* to live the religious life of their choice, carries with it no obligation to take an interest, let alone attribute value to the life they choose. In the best of liberal traditions, *Dignitatis Humanae* calls upon Christians not merely to show tolerance toward members of other faith traditions, but to recognize their fundamental, and religiously sanctioned *right* to their religion. But just as in liberalism, what is solemnly respected and valued is not *the religion they choose,* but only their *right* to choose it.

Nostra Aetate is precisely about that, about the religious value, not of people's right to be religious, but of the various religions they choose; urging Christians to engage in dialogue and collaboration with practitioners of those other religions, and to do so "with prudence and love," so as to "recognize, preserve and promote the good things, spiritual and moral, as well as the socio-cultural values found among these men."

Taken together, *Dignitatis Humanae* and *Nostra Aetate* thus constitute a most positive, dramatic and timely development in Church doctrine, one that with respect to religious pluralism, goes an important step *beyond* liberal democracy; one, to which Judaism and Islam have yet to respond.

But can they? I can only speak for Judaism, of course. Does Judaism offer resources capable of providing a serious, authentic, analogous religious incentive for attributing to other faith traditions sufficient *inherent value* to merit seriously engaging them in study and dialogue? Can one even imagine a practicing orthodox rabbi of standing who is dedicated to studying Christianity with Christians and Islam with Muslims as a solemn religious obligation? Well, far fetched as this may seem, I believe such a Jewish response is not only forthcoming, but capable of going an interesting step beyond the two Roman Catholic documents.

However, what Judaism's formative canon has to offer in this respect, differs significantly, I believe, from the mode of reasoning adopted in the two Catholic documents. Without in any way detracting from their enormous achievement, the approach I shall propose diverges from that of Vatican II in two important respects. First, while *Nostra Aetate* builds upon *Dignitatis Humanae,* urging Christians to take a keen interest in the various ways different people exercise their *right* to their religions, I can see no way of affirming such a right from a Jewish perspective. If a Jewish response to *Nostra Aetate* is at all forthcoming—and I think it is—it must be grounded differently, in a manner that somehow drives a wedge between the two documents: welcoming adherents to other forms of religious and irreligious life as religiously significant partners in dialogue, but without having to acknowledge them as possessing a religiously sanctioned *right* to the lives they have chosen to live.

Second, unlike the universalist religious aspirations of both Christianity and Islam, Judaism's theology of election does not sit at all comfortably with the argument from sameness that grounds *Nostra Aetate*'s call for dialogue and collaboration. A respective Jewish call for such dialogue cannot be grounded on what we might have *in common* with the religious and irreligious other. But if Judaism is incapable of religiously affirming that people possess a fundamental human right to their religions and cultures, as does *Dignitatis Humanae*, and lacks the kind of universalist, humanist presumption of fundamental religious sameness that grounds *Nostra Aetate*, what, in principle *could* serve as a religiously viable justification for such a call? This is the question this paper wishes to engage, and to do so from the traditional perspective of rabbinic Judaism.

To do so I would like to look briefly at rabbinic Judaism's two most important scales of perfection—on the public, communal level, that of the development and perfecting of halakha, religious law, and, on the personal, intra-subjective level, that of the perfecting, or halakhic cleansing of the self. At the high point of each, I shall argue, we encounter the other in a crucially important role.

Perfecting and Purging the Law[2]

The opening sections of tractate *Eduyot*—Testimonies—of the Mishna raises the following question. Why, given that the halakha is always decided by majority vote, do we nonetheless insist on keeping explicit record of the rejected minority opinion? The Mishna offers its question two contrasting answers: presenting the one it adopts as that of the majority, and the one it rejects as that of the dissenting minority voice of one Rabbi Yehuda b. Ilai.[3]

[2] The texts to which this and the next section refer are reproduced below in both the original Hebrew/Aramaic and English translation as appendices A and B respectively.

[3] For a fuller analysis of this and the *Tosefta* text see my discussion of Halbertal's reading of it in M. Fisch, "Canon, controverse et réforme; une réflexion sur l'autre voix du judaïsme talmudique", *Les Cahiers du judaisme*, 18 (2005), 61–75.

One answer asserts that the reason for keeping the rejected minority opinion on the books is to ensure that it *remains* firmly rejected. If in the future a person should claim that the law is thus and so, we will be able to prove to him that his was a minority opinion against which the majority of rabbis had ruled differently. The rejected minority voice is meticulously remembered, if you wish, in order to ensure that it is firmly forgotten!

According to the second answer, it is important to keep good record of the dissenting voices of the tradition just in case future rabbinic authorities will want to overturn their predecessor's majority ruling, and revive the voice rejected. According to this very different view, minority opinion is recorded in order to be kept alive as a viable, revivable halakhic option.

The two answers to the Mishna's seemingly innocent question of why halakhic tracts should record opinions other than the ones they adopt, represent a profound disagreement regarding the very nature of halakhic development. The first approach views halakhic rulings as irrevocable, as fixed forever, and strictly binding on all future generations of halakhists. With regard to all former rulings, halakhic authority remains purely judicial. The legislative authority of even the High Jewish Court of old, the *Sanhedrin*, is firmly limited to new cases with regard to which no existing ruling can be found. And on these questions too, once a vote is taken, the matter will be considered settled and binding for all time.

The second approach views the halakhic authorities of each generation as charged with *legislative*, not merely judicial responsibility, even with respect to existing halakha. It considers it their religious duty not only to transmit and to occasionally supplement the system of law they inherit, but, as true lawmakers should, to hold it in constant review and change whatever they find wanting, overturning former authoritative rulings as they see fit. You can't have it both ways, of course, nor is there a third, intermediate, option—as certain of my critics have tried to suggest over the years.[4] Existing halakhic rulings can be considered the legitimate objects of, or as immune to, later rabbinic criticism and revision, but not as both.

As I said, one of the two views is presented by the Mishna the majority opinion, and hence as the one endorsed as halakha, and the other as that of the minority, and hence rejected voice of R. Yehuda b. Ilai.

---

[4] See for instance the papers by Magid, Cohen and Meskin in the issue of *The Journal for Textual Reasoning* devoted to my *Rational Rabbis: Science and Talmudic Culture*—http://jtr.lib.virginia.edu/volume4/number2/index.html.

Many readers, I am sure, will be surprised to learn that the opinion the Mishna endorses is the reformative approach of halakhic revisionism, while relegating halakhic traditionalism to the rejected minority opinion of Rabbi Yehuda. Astonishingly, however, the ruling of the Tosefta, the tannaitic text, contemporary of the Mishna, which can be found in the right-hand column of the table in Appendix A, is exactly the opposite!—presenting the very same two answers to exactly the same question, while relegating the Mishna's majority ruling to the minority and vice versa!

Let us stay with the Mishna, however, which for the Talmudic literature is by far the more authoritative of the two documents. The Mishna's firm ruling in favor of a reformative view of Jewish law must sound surprising to some readers, coming, as it does, from the very heartland of rabbinic Judaism. So lets bear with the Mishna's ruling a little longer, for, as we shall see, it is only the beginning, for there's a catch.

To see how, let us move on to the next stage of halakhic development immediately following the Mishna, which is most prominently represented by the most important work of early rabbinic Judaism: the Babylonian Talmud—the *Bavli*—and to what I take to be the *Bavli*'s most distinctive, yet misunderstood feature.

In one undisputed sense, perhaps the most important sense, the *Bavli* can be said to fully comply with the Mishna's reformative approach. It is the sense we call in Hebrew מבחן התוצאה, the test of the outcome. The *Bavli* is redacted in the form of often heated disputes between generations of later, amoraic sages, who extensively comment, ponder, deliberate and expound upon each of the Mishna's rulings. What I mean by the test of the outcome is that when one honestly compares the state of halakha the *Bavli* leaves us with with the Mishnaic system from which it sets forth, there can be no doubt that at least *de facto*, the *Bavli*'s decisive contribution to halakha is thoroughly reformative—even dramatically so. Yet, when instead of examining the *Bavli*'s halakhic *conclusions*, one examines the discussions that are purported to yield them, a very different picture seems to emerge. For the *Bavli* hardly ever openly confronts a mishna or any other tannaitic source in the way one would expect legislators who considered themselves authorized to change the law to do. Rather, earlier tannaitic rulings are pondered, analyzed and interpreted as Supreme Justices would deliberate the law—powerless to change it, yet responsible to apply it adequately. However, a closer look reveals that in very many cases—far too many for comfort—the interpretations offered are far-fetched to the point of disbelief. The text marked as Level II in Appendix A presents a simple and wholly commonplace example.

In *Bavli, Sanhedrin*, folio 41a, R. Hisda—a major early 4th century amoraic sage—is reported to have ruled that if under interrogation, the testimonies of witnesses to a murder turn out to contradict each other with respect to the murder weapon—"such as if one testifies that he killed him with a sword, and the other, that he killed him with an axe"—their testimonies cannot be accepted, and the case is dismissed.[5] However, if their testimonies fail to cohere with respect to what the murderer was wearing, for example—"such as if one testifies that he wore white, while the other claims he wore black—their testimonies are to be accepted, and the case heard. An objection to R. Hisda's ruling is then raised by the anonymous narrator of the *sugya* in the form of a tannaitic source that claims explicitly contrary to R. Hisda, that if one witness testifies that the murder weapon was a sword, and the other claims it was an axe, *or* if one testifies that the murderer wore white, and the other that he wore black their testimonies, and with them the case, must be dismissed. To this R. Hisda responds, however, not by reasoning against the source, and explaining why he saw fit to overthrow the former tannaitic ruling and change the law,[6] but by radically reinterpreting the tannatic source as consistent with his position. According to R. Hisda's extremely forced reading, the tannaitic source is read as referring not to the case in which their testimonies differ with regard the murderer's *clothing*, but to the very special case in which they testify that the murderer's garment happened to be the murder weapon, "such as when they testify that he strangled him with his cloak"!

I won't mince my words; R. Hisda's interpretation of the tannaitic text is far-fetched to the point of absurdity. Had the garments mentioned in the ruling's second clause been meant to represent murder weapons, then the entire second clause would have been superfluous. It is not a strained reading, its an *unfounded* one. And what is worrying about it is that such readings are so common in the Bavli's halakhic discourse that seasoned students no longer give them a second thought. Yet, given the logical rigor of that discourse, it is because such feats of quite implausible interpretation *are* so widespread, that one cannot brush them aside as

---

5 In Talmudic law there is no state or county prosecution. Criminal charges are brought before the court by the witnesses to the crime. Court proceedings commence, therefore, with a thorough interrogation of the witnesses by the court. Talmudic law requires a minimum of two (adult male) witnesses whose testimonies both suffice and cohere for a case to be heard.

6 And one has to admit that there are very good reasons for not deeming all testimonial inconsistencies detrimental, unless they pertain to the crime itself.

mere lapses of rationality. What renders the question especially vexing is the fact that far-fetched interpretation seems to make no sense at all from either view of halakhic development and authority mentioned above. For traditionalists,[7] who are committed to the law as they find it, there seems never to be reason for radical reinterpretation, while for anti-traditionalists, who are at liberty to change the law as they see fit, what at all can be achieved by reinterpreting earlier sources?

So what is going on here? I said before that when one compares the halakhic system the *Bavli* leaves us with to that of the Mishna from which its sets forth, it is hard not to characterize its project as thoroughly reformative—which, of course, sits nicely with the Mishna's own reformative position that, as we have seen, urges halakhists to review and revise the system of law they inherit. But there's a catch. According to the Mishna, halakhists can change the law, but they can never *really* reform the system. They can overturn and replace a current ruling, but they cannot *obliterate* it; they can overthrow rulings they find objectionable, but cannot declare them null and void. Rulings they reject cannot be erased, only demoted to the status of minority opinion, where, though not currently in force, will remain forever part and parcel of the halakhic tradition to be studied and transmitted from generation to generation, and, therefore, always liable of being restored. As strongly as halakhists may feel against the former rulings they reject, they are powerless *as legislators* to eliminate them from the record, and render them *unviable*.

It is from this perspective that, I believe, the Bavli's widespread practice of far-fetched, often unfounded halakhic reinterpretation may be made better sense of.[8] For conscientious legislators, the process of halakhic change does not necessarily end with the passing of a new and different

---

[7] "Traditionalist" and "anti-traditionalist" are the terms I introduced in *Rational Rabbis* (pp. 57ff) to describe the two opposing positions. Traditionalists view the halahic authorities as wholly obligated to the halakhic tradition as they find it, and are powerless to change it. Anti-traditionalists, by contrast, view them as obligated to hold the halakhic tradition as they find it in constant review.

[8] For the record, this is my third attempt to date to make sense of the *Bavli*'s widespread practice of far-fetched halakhic interpretation. In *Rational Rabbis* (pt. II, chap. 3) I proposed to view it as a traditionalist surface rhetoric adopted intentionally to conceal from novices the Bavli's reformative agenda. In a later paper, "Forced Readings and Binding Texts: The Amoraic *Ukimta* and the Philosophy of Halakha" (Hebrew), in A. Ravitzki and A. Rosenak (eds.), *New Studies in The Philosophy of Halakha*, the Van Leer Institute and the Magnes Press, Jerusalem, 2008, 311–344, I argued that in attempting to give life to a former text without conceding one's own contrary position can have a self-critical enriching effect on the latter, and is hence in the interest of the anti-traditionalist. The view outlined in the following lines can be seen as either supplementing or replacing those earlier attempts.

ruling. Legislation itself is often only the formal first step in a longer and more significant reform process, in the course of which certain old rulings, norms and halakhic sensibilities do not merely cease *to be the law* by formal acts of law-making, but to cease to be considered as *live halakhic options* by other means. Radical reinterpretation, I suggest, is one such means. By preferring not to openly rule against the tannaitic source, but to reinterpret it in accord with his own more lenient position, later authorities, like R. Hisda, stand a very good chance of achieving what no measure of legislative authority is, in principle, capable of achieving within a reformative setup—namely, the *eradication* of the source's original position from the halakhic record, and hence from the halakhic tradition, and hence from halakhic memory. By arguing against a position, however vehemently, one preserves it and gives it voice; by interpreting it in line with one's own position, one's own position remains the only one on record!

Had space allowed, I would have brought to the table a number of extensive and detailed Talmudic discussions of Mishnaic bodies of law, in which, under the guise of innocent pondering, the Mishna is so radically transformed by reinterpretation as to effect major halakhic revolutions.[9] To paraphrase the name of the famous movie, in this way, time and time again the Mishna's original positions are deliberately 'lost in interpretation', as it were, lost from the halakhic record, purged from the halakhic tradition, no longer preserved even as a minority view.

This, I believe, is how the Talmud performs the lion's-share of its halakhic work. But sometimes even harsher methods are needed. Some, particularly blatant halakhic rulings do not lend themselves so easily to radical reinterpretation. I'm referring to rulings that are considered not merely mistaken or wrongheaded or ineffective, but downright offensive—which brings me to the third and highest rung on the scale of halakhic perfection, and with it, just in case you were beginning to wonder, the main point of this long section. These are cases in which a method of purging even more radical than radical reinterpretation is called for.

The first of the three texts comprising Level III of Appendix A offers a brief yet potent example. The Jerusalem Talmud, the *Yerushalmi*, relates the following story about the great sage of old, R. Simeon b. Shatah, who

---

[9] For one particularly vivid example (though presented along lines different from those proposed here) see A. Furstenburg, "Restitution of Lost Property in the Tannaitic and Amoraic *Halakhah*: A Preliminary Philosophical Study of the Forming of a Conception" (Hebew), *AJS Review*, 29:1 (2005): 1–51.

we are told, had to work extremely hard in the cotton business to make a living. His students decided to purchase him a donkey which he could then hire out and be able to devote more time to his studies. They did so, only to find to their surprise that the Syrian from whom they had bought it had not noticed a precious gem lodged in its harness. Now, Master, they announced joyfully you will no longer have to work at all. "Did the owner know he was selling the gem with the donkey?" enquired R. Simeon "No," they answered, "of course not." "Then return it to him immediately!" he commanded. "But one is permitted by law to keep it," the Talmud says—for according to halakha, explains the narrator, a Jew is not obliged to return the lost property of a gentile, only that of another Jew. "Do you take R. Simeon for a barbarian?!" stormed Rabbi Yehuda the Prince, putting an abrupt end to the conversation.

Now, what is fascinating about this little, yet most significant tale, strategically placed at the heart of the *Yerushalmi*'s discussion of the laws of lost property, is the way an objectionable, yet bona fide halakhic ruling is not questioned, nor reinterpreted, nor legislated against—all of which would have left it on the books as a viable option of some sort. It is not even hailed a mistake. It is left on the books as the law, its obnoxious original meaning intact for all to see, but is declared barbarous—a law indeed, but a law only a barbarian would follow!

In the second text, Maimonides, the great halakhist and codifier of the 13th century, presents the same kind of move in his great halakhic work in stark oxymoronic formulation: It is permitted to work a gentile slave with rigor, he rules, yet those who do so display the kind of cruelty and aggression found only among heathen idolaters. Again, as with R. Simeon and his students, the law that permits Jews to work their non-Jewish slaves rigorously is declared impious, unwise, unmerciful and unjust—a law befitting of idolaters. Here again, the objectionable halakha is not ruled against or reinterpreted. It is left on record as the law, but is rendered ineffective and as effectively as possible unrevivable, by being utterly morally stained.

Over and above the keen reformative work of legislators and interpreters, the law is thus also kept in check by subjection to the kind of moral, extra-halakhic critique that we have just witnessed. But by what code and according to which values is the critique to be leveled? How does the tradition gauge what is to be considered barbarous, unjust, unwise, unmerciful, and what is to be considered the will of God, by what, and by whose ethical code? The giveaway phrase can be found in Rabbi Yehuda the Prince's very last statement appended to the story of R. Simeon and his donkey in the name of Rabbi Yehudah the Prince. The great sage refused to keep the gem, and declared doing so barbarous, he explains, because he "was willing to relinquish all the riches of the world *just to hear people declare 'blessed be He God of the Jews'*." The moral standard against which halakha is to be measured is that of the gentile not necessarily that of the Jew. In developing halakha, and especially in the moral critique of existing halakha, the halakhist's first obligation is to what, in a different context, as we shall see in a moment, the Talmud refers to as to "Sanctify God's name"—the term it uses in that context to mean, being forever open to the world, attentive to how our laws, our normative choices, and conduct, appear, sound and morally resonate outside the boundaries of the halakhic community. To Sanctify God's name is to transcend the all-too-cozy, all-too-self-righteous and self-congratulating dialogue conducted by Jewish halakhists among themselves, not by vainly claiming to speak in the name of God's attributes or in the name of the one true timeless moral code of the Torah, but by attempting far more modestly, to create an inner distance from what we take for granted by prudently attempting to see ourselves as others do, and to try prudently to live up to the standards of the rest of the civilized world as we find it. This is because others are capable of giving us what we are in principle unable to give ourselves: a truly critical perspective.

The third and last of the Level III texts makes the point very nicely in a different and perhaps more subtle way. The Law of Damages states that the owners are responsible to compensate for injuries inflicted by their oxen on other oxen. However, in the case of a first offender, as it were, the ox is regarded "innocent" and its owner is charged for only half the damage. Owners of oxen with a history of violence, are expected to be more cautious, and in the case of a goring must pay the damage in full. However, this applies only to Jews. "Where an ox belonging to an Israelite has gored an ox belonging to a Canaanite," rules the Mishna, "there is no liability, whereas where an ox belonging to a Canaanite gores an ox belonging to an Israelite, whether *tam* (innocent) or *mu'ad* (with a history), the compensation is to be made in full." How can this be?—asks the Talmud. Either the Law of Damages applies to gentiles or it does not. If it does, they should be compensated in the case of injury to their oxen like anyone else, and if it does not, they should not be required to compensate others when their oxen are to blame. But you cannot have it both ways. The Talmud offers two alternative explanations. The law, both admit, is indeed blatantly unfair to non-Jews, but the gentiles deserve to be treated thus, because they were offered the Torah and refused it, claims one; because of their failure to abide even by the seven Noachite laws, claims the other. And at this point, without any forewarning, comes the story of the two benign legal experts sent from Rome:

> Our Rabbis taught: The Government of Rome had long ago sent two commissioners to the Sages of Israel with a request to teach them the Torah. They read it [the Torah] once, twice and thrice. Before taking leave they made the following remark: We have gone through your Torah carefully, and have found it correct with the exception of this point, viz. your saying that if an ox of an Israelite gores an ox of a Canaanite there is no liability, whereas if the ox of a Canaanite gores the ox of an Israelite, whether *Tam* or *Mu'ad*, compensation has to be paid in full. In no case can this he right. For if [the law] is meant to imply only to 'his neighbour' [i.e. to Jews], why then in the case of an ox of a Canaanite goring an ox of an Israelite should there also not be exemption? If [on the other hand] it was not meant to apply only to 'his neighbour', why then even in the case of an ox of an Israelite goring an ox of a Canaanite, should there not be liability? We will, however, not report this matter to the authorities [in Rome].

This is a wonderful Talmudic moment, in which, by means of a story inserted into the legal discussion, the student and reader are allowed a serious peek at the how the system they are committed to and committed to defend and justify, looks like from the point of view of a friendly, civilized outsider. And the picture isn't at all pretty. What renders this particular Talmudic moment truly fascinating is the way the story is left to hover and to cast its dark and worrying shadow over the self-congratulating discussion, without need for another word.

The Scale of Self Purging

The final text, Appendix B, really needs no introduction. It vividly describes the various levels of, and means for, personal, deontic self-cleansing. It is a scale culminating in the very gravest of religious offences, worse than murder, worse than adultery, worse than idolatry—the profanation of the Name, *hilul ha-Shem*—set up, *sui generis*, as a category in itself, set far above all other capital punishment offences. Yet intriguingly, what constitutes the ultimate transgression is not determined by the Torah code it transgresses. Not at all. Unlike all others, the gravest transgression of all is not defined absolutely but relatively; relative to how one's conduct seems—and again, not to how it seems to Torah experts, but to those outside the system looking in!

There exists, then, a central Talmudic voice, that full-consciously locates the definitive corrective, the ultimate normative check on both

Judaism's system of law and a Jew's personal conduct, in the moral critique leveled at them by the civilized other. Self-critique, especially normative self-critique is severely limited, especially with regard to basics.[10] And it is here, in the moral challenge leveled at us by adherents to ways of life very different from, even contrary to our own, that the Talmud locates the ultimate source of moral correction. Here lies, I believe, the basis for an authentic Jewish welcoming response to Vatican II.

A real-life example may help better explain what I am trying to get at. Some years ago the Israeli daily *Haaretz* published a detailed profile and long interview with an ultra-orthodox American women who works as a "rabbinic pleader," i.e., a barrister to the rabbinical courts, in New York City. The interview focused on the built-in halakhic biases against women, and the ways of legally reasoning round them in the rabbinic courts. At one point, in the course of addressing various gender-based asymmetries with regard to rabbinic conceptions of marriage and divorce, she noted emphatically how, due to the growing number of same-sex couples in the city, we have come to realize how serious couples can maintain perfectly equal and symmetrical relationships of a kind we formerly did not think possible! Now, this is not an argument in favor of homosexuality, the gay community, or same-sex marriage legislation. In fact, I am sure that as the ultra-orthodox person she is, she most probably considers them abominations. Nonetheless, utterly sinful and religiously loathsome as she may take it to be, there was something vitally important *religiously* that she admits the gay community was able to teach her, which had it not existed, she would never have learned! Homosexuality remains for her a terrible sin, for which nobody could be considered as having a religiously sanctioned *right* to commit. Yet, thank God for same-sex marriages!—she all but exclaims, for they challenge us by their very otherness, or sameness, I should say. In other words, although she could never apply to the gay way of life anything akin to *Dignitatis Humanae*, she can, and certainly does regard it in a manner closely akin to *Nostra Aetate*—denouncing it as profoundly sinful, while learning from it a profoundly valuable religious lesson; a way of life wholly rejected, while deeming its challenging presence a rare and unique blessing!

---

10 I have devoted a book-length study to exploring the limits of normative self criticism, and the crucial role of external normative critique in this respect. See M. Fisch and Y. Benbaji, *The View from Within; Normativity and the Limits of Self-Criticism*, South Bend IN: The University of Notre Dame Press, 2011. See also my "Science, Religion, and Rationality—a Neo-Hegelian Approach", *The Toronto Journal of Theology* 29, no. 2 (2013): 319–336 (also reprinted above, pp. 115–134).

Are there authentic resources for such a double standard in the rabbinical tradition? In the light of the texts I have presented and many others, I sincerely believe that there are. For as we have seen, they bespeak profound acknowledgment of the vital and essentially *religious* need for the thoughtful external critique of those whose normative commitments differ significantly from our own, as a constant check on our habitual ways of thinking. Maimonides's account of *T'shuva*, repentence, could have joined Simeon ben Shatah's purchased donkey, the Roman legal experts' assessment of the halakhic Law of Damages, and the outward-looking, essentially relativistic criterion of *hilul ha-Shem*—the desecration of God's name—to bear vivid witness to such an approach. Only in the context of the trusted normative criticism of others is one capable of gaining a critical perspective on the self-congratulating and self-justifying hold of one's own normative commitments, all these interesting passages clearly imply. These are the Jewish resources I would build on not only for engaging *Nostra Aetate*, but for going an important step beyond the firm inter-religious boundaries to which, because coupled to *Dignitatis Humanae, Nostra Aetate* is made to apply.

Let me end this essay with the words with which I concluded another presentation, one delivered from the podium of the Senate hall of the Gregorian University in Rome, at the opening session of an unprecedented conference organized by the Vatican, that represented like no other, the Church's new commitment to serious religious dialogue with Judaism. The Conference was entitled "The First Roman Consultation on Jewish and Canon Law" and was devoted to "Norm Making and Norm Changing" in the two legal systems. I ended that address, as I did here, by alluding briefly to the *Yoma* text. "In the light of this", I said there,

> participating in this conference acquires, from my orthodox Jewish perspective a special religious significance. For there is an important sense in which, by engaging in serious open and critical inter-faith dialogue... with as significant and important an other as the Catholic Church, that the name of God is sanctified. It is sanctified not because we purport to bring to Rome His one true word, but because there is a very good chance that we shall be returning home from Rome better equipped for the ongoing task of working it out for ourselves.[11]

[11] M. Fisch, "Legislation and Legal Norm Development in Jewish Law", *Periodica de re Caronica*, 96 (2007), 401–17, at p. 417.

Appendix A: Three Levels of Halakhic Perfection

Level I: Reform by Argument—The Testimony of *Eduyot*

משנה, עדויות פרק א

ולמה מזכירין דברי היחיד בין המרובין. הואיל ואין הלכה אלא כדברי המרובין? **שאם יראה בית דין את דברי היחיד ויסמוך עליו**....

א״ר יהודה אם כן למה מזכירין דברי היחיד בין המרובים לבטלן? **שאם יאמר האדם כך אני מקובל יאמרו לו כדברי איש פלוני שמעת.**

תוספתא, עדויות פרק א

לעולם הלכה כדברי המרובין. לא הוזכרו דברי יחיד בין המרובין אלא **שמא הוצרך להן שעה ויסמכו עליהן**— דברי רבי יהודה.

וחכמים אומרים: לא הוזכרו דברי יחיד בין המרובין **אלא מתוך שזה אומר טהור וזה אומר טמא כדברי ר׳ אליעזר, יאמרו לו: כדברי ר׳ אליעזר שמעת.**

Mishna *Eduyot* 1

And why do they record the opinion of a single person among the many, when the halakha must be according to the opinion of the many?

[Say the Sages] **So that if a *Bet din* prefers the opinion of the single person it may depend on him**....

R. Yehuda says: if so why are the words of the one (minority) mentioned together with those of the many (only) to dismiss them? **In order that if a person should say such (and such) is my received view, he will (can) be told: you have heard the words of so and so**.

Tosefta *Eduyot* 1

The halakha is always in accord with the opinion of the many. (If so) why do they record the opinion of a single person among the many just in order to set them aside?

R. Yehuda says, they record the opinion of a single person among the many **to enable a Bet din that prefers the opinion of the single person to depend upon it**.

But the sages say, they record the opinion of a single person among the many but **for the sake of cases where this one will rule (a thing) pure and that one will rule (the thing) impure, as did (for example) R. Eliezer. For he can then be told that he had received the words of R. Eliezer**.

Level II: Reform by Rereading—Bavli, *Sanhedrin* 41a

אמר רב חסדא: אחד אומר, בסייף הרגו ואחד אומר באירין [בגרזן] הרגו—אין זה נכון. אחד אומר כליו שחורין, ואחד אומר כליו לבנים—הרי זה נכון. מיתיבי:...אחד אומר בסייף הרגו ואחד אומר באירין הרגו, אחד אומר, כליו שחורין ואחד אומר כליו לבנים—אין זה נכון! תרגמה רב חסדא: בסודר שחנקו בו...תא שמע: אחד אומר סנדליו שחורין ואחד אומר סנדליו לבנים—אין זה נכון! התם נמי, כגון שבעט בו בסנדלו והרגו.

Rav Hisda said: If one testifies that he killed him with a sword, and one, that he killed him with an axe, [their testimony] is unacceptable. [But] if one testifies that he wore black and one that he wore white, it is acceptable.

[An] objection [was raised from a tannaitic source]…"If one testifies that he killed him with a sword, and one, that he killed him with an axe, [or] if one testifies that he wore black and one that he wore white, it is unacceptable." Rav Hisda interpreted [the tannaitic source] as [referring to cases in which] he was strangled by the garment…

Come hear [yet another tannatitic source]: "If one testifies that his sandals were black, and one, that they were white, it is unacceptable". There too [the source refers, according to Rav Hisda] to such cases as if he kicked him with his sandal and killed him.

Level III: Reform by Moral Staining

1. Yerushalmi, *Bava Metzia* 8c

שמעון בן שטח הוה עסיק בהדא כיתנא. אמרין ליה תלמידוי: ר׳ ארפי מינך ואנן זבנין לך חדא חמר ולית את לעי סוגין." אזלון זבנון ליה חדא חמר מחד סירקאי, ותלי ביה חדא מרגלי. אתון לגביה אמרין ליה: מן כדון לית את צרי' לעי תובן! אמ׳ לון: למה? אמרין ליה: זבנינן לך חד חמר מחד סירקיי ותלי ביה חדא מרגלי. אמ׳ לון: וידע בה מרה? אמרין ליה לא. א"ל: לון איזל חזר. לא כן אמר רב חונא...אפילו כמאן דמר גזילו של גוי אסור כל עמא מודיי שאבידתו מותרת? **מה אתון סברין שמעון בן שטח ברברון הוה?! בעי הוה שמעון בן שטח משמע "בריך אלההון דיהודאי" מאגר כל הדין עלמא!**

Simeon b. Shatah made a living in the cotton business. His students said to him, ease off, we shall purchase for you a mule so that you won't have to work so much. They went and purchased him a mule from a certain Syrian, and [subsequently] found a precious stone hanging from it[s neck]. They came to [R. Simeon] and said, from now on you needn't work at all! Why?, he asked. Because we purchased you a mule from a certain Syrian, on which a precious stone was hanging! And did he [the owner] know of it? He asked. No, they answered. Go return it to him! He ordered. But [they asked], doesn't R. Huna…rule that even those who claim that one may not steal from a gentile, admit that one is not obliged to return a gentile's lost property [as one is obliged to return that of a Jew]?

Do you take Simeon son of Shatah for a barbarian?!

Simeon b. Shatah [the narrator explains] **would rather hear "Blessed is the God of the Jews" than receive all the riches in the world!**

2. Maimonides: Laws of Slavery 9:7–8

מותר לעבוד בעבד כנעני בפרך, ואע״פ שהדין כך מדת חסידות ודרכי חכמה שיהיה אדם רחמן ורודף צדק ולא יכביד עולו על עבדו ולא יצר לו ויאכילהו וישקהו מכל מאכל ומכל משתה.... וכן לא יבזהו ביד ולא בדברים—לעבדות מסרן הכתוב לא לבושה. ולא ירבה עליו צעקה וכעס אלא ידבר עמו בנחת וישמע טענותיו....ואין האכזריות והעזות מצויה אלא בעכו״ם עובדי ע״ז אבל זרעו של אברהם אבינו והם ישראל שהשפיע להם הקב״ה טובת התורה וצוה אותם בחקים ומשפטים צדיקים רחמנים הם על הכל, **וכן במדותיו של הקב״ה שצונו להדמות בהם** הוא אומר ״ורחמיו על כל מעשיו״.

It is permitted to work a gentile slave with rigor.

However, although such is the law, it is the quality of piety and the way of wisdom that a man be merciful and pursue justice and not make his yoke heavy upon his slave or distress him, but give him to eat and to drink of all foods and drinks....

Nor should the master heap upon the slave physical or oral abuse and anger, but should rather speak to him softly and listen to his claims....

Cruelty and aggression are found among heathen idolaters, but the descendants of Abraham, The People of Israel upon whom God has bestowed the benefits of Torah with its just and righteous statutes and laws, that are merciful to all, and in this way emulate God's own attributes, as it says "And His mercy is on all His works" (Psalms 145:9)

3. Bavli, *Bava Kama* 37b–38a

מתני׳....שור של ישראל שנגח לשור של כנעני—פטור, ושל כנעני שנגח לשור של ישראל, בין תם בין מועד—משלם נזק שלם....

גמ׳...שור של ישראל שנגח שור של כנעני—פטור. אמרי: ממה נפשך? אי רעהו דוקא, דכנעני כי נגח דישראל נמי ליפטר! ואי רעהו לאו דוקא, אפילו דישראל כי נגח דכנעני נחייב!

א״ר אבהו...

רבי יוחנן אמר...

וכבר שלחה מלכות רומי שני סרדיוטות אצל חכמי ישראל למדונו תורתכם, קראו ושנו ושלשו. בשעת פטירתן, אמרו להם: דקדקנו בכל תורתכם ואמת הוא, חוץ מדבר זה שאתם אומרים: שור של ישראל שנגח שור של כנעני—פטור, של כנעני

שנגח שור של ישראל, בין תם בין מועד—משלם נזק שלם, ממ"נ? אי רעהו דוקא, אפילו דכנעני כי נגח דישראל ליפטר! ואי רעהו לאו דוקא, אפילו דישראל כי נגח דכנעני לחייב! ודבר זה אין אנו מודיעים אותו למלכות.

Mishna:... Where an ox belonging to an Israelite has gored an ox belonging to a Canaanite, there is no liability, whereas where an ox belonging to a Canaanite gores an ox belonging to an Israelite, whether while tam or mu'ad, the compensation is to be made in full.

Gemarah:... But I might here assert that you are on the horns of a dilemma. If the implication of 'his neighbour' has to be insisted upon, then in the case of an ox of a Canaanite goring an ox of an Israelite, should there also not be exemption? If [on the other hand] the implication of 'his neighbour' has not to be insisted upon, why then even in the case of an ox of an Israelite goring an ox of a Canaanite, should there not be liability!?

R Abbahu said:...

R. Johanan said...

... Our Rabbis taught: The Government of Rome had long ago sent two commissioners to the Sages of Israel with a request to teach them the Torah. They read it [the Torah] once, twice and thrice. Before taking leave they made the following remark: We have gone carefully through your Torah, and found it correct with the exception of this point, viz. your saying that if an ox of an Israelite gores an ox of a Canaanite there is no liability, whereas if the ox of a Canaanite gores the ox of an Israelite, whether Tam or Mu'ad, compensation has to be paid in full. In no case can this he right. For if it is meant to imply only to 'his neighbour', why then in the case of an ox of a Canaanite goring an ox of an Israelite should there also not be exemption? If [on the other hand] it was not meant to apply only to 'his neighbour', why then even in the case of an ox of an Israelite goring an ox of a Canaanite, should there not be liability? We will, however, not report this matter to the authorities [in Rome]

Appendix B. The Scale of Self-Perfection

Bavli, Yoma 86a

שאל רבי מתיא בן חרש את רבי אלעזר בן עזריה ברומי: שמעת ארבעה חלוקי כפרה שהיה רבי ישמעאל דורש? אמר: שלשה הן, ותשובה עם כל אחד ואחד.

עבר על עשה ושב—אינו זז משם עד שמוחלין לו...

עבר על לא תעשה ועשה תשובה—תשובה תולה, ויום הכפורים מכפר שנאמר....

עבר על כריתות ומיתות בית דין ועשה תשובה—תשובה ויום הכפורים תולין, ויסורין ממרקין....
אבל מי שיש חילול השם בידו—אין לו כח בתשובה לתלות, ולא ביום הכפורים לכפר, ולא ביסורין למרק. אלא כולן תולין, ומיתה ממרקת, שנאמר (ישעיהו כב) "ונגלה באזני ה׳ צבאות אם יכפר העון הזה לכם עד תמתון".
היכי דמי חילול השם? אמר רב: כגון אנא. אי שקילנא בישרא מטבחא ולא יהיבנא דמי לאלתר....
רבי יוחנן אמר: כגון אנא דמסגינא ארבע אמות בלא תורה ובלא תפילין.
יצחק דבי רבי ינאי אמר: כל שחביריו מתביישין מחמת שמועתו (היינו חילול השם)...
אביי אמר: כדתניא, (דברים ו) "ואהבת את ה׳ אלהיך"—שיהא שם שמים מתאהב על ידך, שיהא קורא ושונה ומשמש תלמידי חכמים, ויהא משאו ומתנו בנחת עם הבריות, מה הבריות אומרות עליו—אשרי אביו שלמדו תורה; אשרי רבו שלמדו תורה; אוי להם לבריות שלא למדו תורה; פלוני שלמדו תורה—ראו כמה נאים דרכיו; כמה מתוקנים מעשיו; עליו הכתוב אומר (ישעיהו מט) "ויאמר לי עבדי אתה ישראל אשר בך אתפאר". אבל מי שקורא ושונה ומשמש תלמידי חכמים ואין משאו ומתנו באמונה, ואין דבורו בנחת עם הבריות, מה הבריות אומרות עליו—אוי לו לפלוני שלמד תורה; אוי לו לאביו שלמדו תורה; אוי לו לרבו שלמדו תורה; פלוני שלמד תורה—ראו כמה מקולקלין מעשיו וכמה מכוערין דרכיו! ועליו הכתוב אומר (יחזקאל לו) באמר להם עם ה׳ אלה ומארצו יצאו.

R. Matthia b. Heresh asked R. Eleazar b. Azariah in Rome: have you heard about the four ranks of forgiveness, concerning which R. Ishmael has lectured? He answered: They are three, and with each of them repentance is connected—If one transgressed a positive commandment, and repented, then he is forgiven, before he has moved from his place;... If he has transgressed a prohibition and repented, then repentance [can only] suspend [the punishment] and the Day of Atonement procures atonement;... If he has committed [a sin to be punished with] extirpation or death through the Beth Din, and repented, then repentance and the Day of Atonement [can only] suspend [the punishment thereon], and suffering completes the atonement;... But if he has been guilty **of the profanation of the Name,** then penitence has no power to suspend punishment, nor the Day of Atonement to procure atonement, nor suffering to finish it, but all of them together [suffice only to] suspend the punishment and only death finishes it;...

What constitutes profanation of the Name?—Rab said: If, e.g., I take meat for the butcher and do not pay him at once.... R. Johanan said: In my case [it is a profanation if] I walk four cubits without [uttering words of] Torah or [wearing] tefillin.

Isaac, of the School of R. Jannai. said: If one's colleagues are ashamed of his reputation, that constitutes a profanation of the Name....

Abaye explained: As it was taught: And thou shalt love the Lord thy God (Deut. 6) i.e., that **you should cause the Name of Heaven be beloved.** If someone who studies and restudies Torah, and attends on the disciples of the wise, is honest in business, and speaks pleasantly to persons, what do people then say of him? 'Happy is the father who taught him Torah, happy is the teacher who taught him Torah; woe unto people who have not studied the Torah; for this man has studied the Torah look how fine his ways are, how righteous his deeds! Of him does Scripture say: And He said unto me: Thou art My servant, Israel, in, whom I will be glorified. (Is. 49) But if someone studies and restudies Torah, attends on the disciples of the wise, but is dishonest in business, and discourteous in his relations with people, what do people say about him? 'Woe unto him who studied the Torah, woe unto his father who taught him Torah; woe unto his teacher who taught him Torah!' This man studied the Torah: Look, how corrupt are his deeds, how ugly his ways; of him Scripture says: "In that men said of them: These are the people of the Lord, and are gone forth out of His land" (Ez. 36).

INTERVIEW WITH MENACHEM FISCH
JUNE 15, 2014

Hava Tirosh-Samuelson

Professor Fisch, thank you for agreeing to participate in the Library of Contemporary Jewish Philosophers. The interview will begin by discussing your upbringing, your intellectual trajectory, and your intellectual identity. Please tell us where were you born and where did you spend your formative years? What kind of education did you receive?

I was born in Leeds, England, to Harold and Joyce Fisch, both of whom were born in England, too. My grandfather on my father's side was Rabbi Solomon Fisch, who in a sense set the tone for my father and even for myself. He was born in Wolbrom, Poland, left home at a young age to study Torah in Germany, and finally migrated to England to serve as the rabbi of small communities in Liverpool, Birmingham, Sheffield, and eventually Leeds. He was a strict, old-school, Orthodox community rabbi. Yet at the same time, he studied for a doctorate in divinity at Manchester University. His doctoral dissertation, which took him years to complete, was a scientific edition of *Midrash Hagadol Bamidbar*, a late rabbinic midrashic text. The critical edition he produced was based on a Yemenite manuscript, for which he had to master Arabic and Greek to address the many philological problems it posed. So already my grandfather possessed a sort of dual personality: a serious religious commitment on the one hand and a deep commitment to academic research on the other. But it wasn't a split personality. My grandfather's academic and scientific involvement was an integral part of his religious persona. He was the model for my father, Harold Fisch, who was also a deeply religious Jew, but at the same time, a first rate scholar, fully at home in Western culture.

After World War II, my father, who had already earned a B.A. in English literature, went to Oxford to complete his graduate work, and he was handpicked by the provost of Leeds University, together with Asa Briggs and Steven Toulmin, to serve on its faculty. At the time he was twenty-three or twenty-four and did not even have a doctorate. My parents made their first home in Leeds, where I was born in 1948. After ten years at Leeds university, and critical of the way the British had conducted themselves during the Suez crisis (1956), my father accepted an invitation from the newly founded Bar-Ilan University in Israel to help establish the English Department there. So in the summer of 1957, just as I was turning nine, the family made aliyah and the rest, as it were, is history.

I would like you to say more about your father's association with Bar-Ilan University. Did he not eventually become the Rector of the university?

Yes. For my father Bar-Ilan was a dream come true, a fully fledged university he saw as dedicated to the same authentic twofold commitment to religious seriousness and academic excellence that he had experienced at home.

What was the impact of Bar-Ilan University on your upbringing and intellectual makeup?

I did my first degree at Bar-Ilan in physics and mathematics. My high school education was an interesting experience, and my years there were undoubtedly formative in certain respects, but the school was not very good. Midrashiyat Noam, of which you may have heard, was one of the first two so-called yeshiva high schools that were founded in Israel during the mid-1940s. However, neither of them succeeded as did Maimonides, the high school established a few years earlier by Rabbi Joseph Dov Soloveitchik and his wife in Boston, in offering an authentic and natural welding of a traditional religious education and a Western secular education. Maimonides School modeled itself on the persona of Rabbi Soloveitchik and translated his philosophy into a way of life. In this environment being an Orthodox Jew and being committed to philosophy were complementary aspects of an integrated and religiously engaged life. In Israel, by contrast, the yeshiva high schools were far more schizophrenic. The two halves of the day (i.e., the mornings devoted to traditional Jewish learning and the afternoons to secular learning) failed to mesh and to relate in any way to one another. In fact, the secular subjects were implicitly taught as merely instrumental to getting ahead in society. This was a school for the so-called religious intelligentsia in Israel, but the school never really took itself seriously as pioneering a new and truly integrated vision of Torah and *derech eretz*. Indeed, unlike Rabbi Soloveitchik, the two rabbis who established the two first yeshiva high schools in Israel, Rabbis Neriah at Kfar Haroeh and Yagel—or Yogel as we all called him—at the Midrashiya, were Haredi-trained rabbis with little if any personal interest in secular learning. With them as role models, it is not surprising that in the schools they established the value of secular learning was more instrumental than an authentic integrated vision of Torah.

The instrumental value of secular learning is just one side of the story. The other side of my high school education (which is still true of many Orthodox high schools today) is that the sciences were considered relatively safe for Orthodox Jews, while the humanities were perceived as posing a threat to religion. The implicit message we received was that if we were determined to pursue academic studies, they should be in the sciences. As a result, upon leaving the Midrashiya, I enrolled in a B.Sc. in physics and mathematics at Bar-Ilan. And it took the entire duration of my undergraduate degree (which in those days was four years) and well into my M.Sc. (working with Prof. Max Jammer on the various proofs of the CPT theorem) for me to realize that I had no desire to become a physicist, and the best of another six years in the army to realize what I did want to do.

That was a high price to pay!

It was, although one can never know if one's formative years would have been better catered for elsewhere. In retrospect, there's a sense in which the ambivalent zigzagging of my early years can be said to have paid off.

Did you choose theoretical physics and mathematics because these subjects were religiously "safe"?

Of course not, it would take me many years to come to that conclusion. At the time I was totally unaware of the intellectual significance or implications of my choice of study. In retrospect, I believe I was tacitly nudged in that direction by the school, which had less trouble engaging modernity by means of science and technology than by modern philosophy, literature, or art. Indeed, to this very day, exceedingly few Israeli yeshiva high school graduates have made their mark in the humanities (except in Jewish studies, of course). And of course, my deeply religious father's unabashed career in English literature "helped" considerably muffle the Midrashiya's tacit message.

And what about Bar-Ilan University? Once you enrolled at the university, did you find a better integration of Judaism and secular knowledge?

Yes, I think Bar-Ilan's commitment to Torah and *derech eretz* presented a far clearer and principled commitment to the integration of the two worlds than the Midrashiya or any other yeshiva high school ever achieved.

To my knowledge the Torah and *Avodah* ideal has something to do with the ideology of socialist Zionism and there was a close connection between the Torah and *Avodah* ideal and the ideology of the labor movement in pre-state Israel.

Yes, that is correct. The Torah and *Avodah* movement started there, but the phrase became shorthand for having one foot in the "tradition" and one foot in the "world." And when it comes to a university, having a foot in the world meant having a foot in the academic world. Bar-Ilan viewed itself as manifesting the Torah and *Avodah* ideal on the academic level.

What about the early beginning of the Torah and *Avodah* movement in Germany? Didn't the movement reflect religious socialism in Germany, which was transported to pre-state Israel?

Yes, but once the movement was transplanted here, it took on this broader meaning of an anti-Haredi attitude of combining this-worldliness with Jewish religious life. Because Bar-Ilan viewed itself first and foremost as a university, it succeeded in forging a far more consistent integration of Judaism and academe than the yeshiva high schools, even though both institutions viewed themselves as rallying under the slogan of "Torah and *derech eretz*." The "and" received different meanings in the two contexts. While the high school used the rhetoric of Torah and *derech eretz*, in reality the curriculum was deeply divided between the religious and the secular spheres. As students we lived looking up to people like Rabbi Neriah and Rabbi Yogel, whose lives were even less integrated than the schools that they founded. It took me a long time to recombine the two spheres of my life and sort out the intellectual relationship for me between traditional Judaism and science.

And Bar-Ilan?

Bar-Ilan University, by contrast, made a real effort to integrate Orthodox Judaism and academe. One of my father's great disappointments was that the university never succeeded in establishing itself as the natural continuation of the yeshiva high schools in the eyes of religious Zionism. Rather, religious Zionism established a whole array of high yeshivot, that is, institutions of purely traditional advanced Jewish learning that were no longer committed to the academic ideal of Torah and *derech eretz*. In this respect religious Zionism has come full-circle intellectually, and is now exclusively led by a growing and extremely influential cadre of rabbis who have no secular education to speak of, dedicated to luring all gifted graduates of the religious high school system into their yeshivot, rather than view Bar-Ilan's initial integrated vision as their ideal.

So there's an interesting and troubling story to tell about Israel's attempt to create an educational system similar to that of Maimonides in Boston and its natural continuation in Yeshiva University on a very large scale, which, for lack of rabbinical leaders like Rabbi Soloveitchik, who viewed his studies at the University of Berlin as part and parcel of his rabbinical formation, has so far failed to properly implement the integrative ideal to which it supposedly stood for.

Such a natural combining of the two worlds is precisely what inspired my grandfather's twofold lifework that my father adopted and eventually attempted to help re-create institutionally at Bar-Ilan. He set out with high hopes of seeing it through, and dedicated his life to doing so, but by the end of the day, he and his Bar-Ilan colleagues were largely unsuccessful.

As an institution, Bar-Ilan remains at least rhetorically committed to its original vision (although it is often criticized for not living up to it, especially in Jewish studies) but in the eyes of religious Zionism, the university is perceived today more as a compromise than as the natural culmination point of its educational vision, and this has had a corroding effect on the university itself.

I am glad you used the phrase "religious Zionism." Do you identify yourself as a "religious Zionist?" If not, what kind of phrase best fits you?

I'm certainly religious, and I'm certainly Zionist. But I'm not a religious Zionist in the normal sense of the term.

If you're going to choose a phrase that best describes you, what would that be?

Do you want me to describe myself?

Yes, please.

Well, you don't ask a philosopher for a phrase, and especially if the label is a kind of an insult.

It is very interesting that the phrase "religious Zionist" has become so unpopular or even toxic. You are not the first person I met who rejects the label even though the person is both religious and a Zionist.

Yes, there is a reluctance to use the phrase "religious Zionist" today because that label has been hijacked in a particular direction and is associated with a particular ideology and a particular political program. Worse, it now connotes a particular and particularly worrying political conception of what Israel is and what she should stand for. But as I see it, the current usage of "religious Zionism" is a misappropriation of and a departure from the original meaning of the term. This is not merely an ideological or political disagreement. I think it goes to the very heart of the revolutionary nature of political Zionism, which traditional Judaism believed and still believes it had yearned and prayed for for millennia, but never really did, and still lacks the basic vocabulary to live, describe, and identify with it.

This is a crucial issue that we will continue to explore throughout the interview. For now let us continue to explore the education you received in Bar-Ilan University. That institution offered you a way to integrate your religious life and modernity.

Yes, that is correct.

But you did not remain in Bar-Ilan for the rest of your academic career. Rather, you moved to Tel Aviv University, a decidedly secular institution, where you teach in the Philosophy Department as well as in the Cohn Institute for History and Philosophy of Science.

Yes, I'm the proud first Ph.D. graduate of the Cohn Institute at Tel Aviv, where I now hold the Joseph and Ceil Maser Chair for History and Philosophy of Science.

Isn't that the chair that the late Amos Funkenstein, cofounder of the Institute, held until his death in 1995?

Yes, the chair was originally established for Amos.

Since you teach philosophy of science, I would like to know more about your philosophic training. Which philosophical traditions did you study? Why did you choose to go into history and philosophy of science? After all, this is not an ordinary decision for a person who studied at Bar-Ilan University or for a religious Jew committed to the study of Jewish philosophy.

Going into history and philosophy of science was part accident and part not. As I said before, I studied physics and mathematics at Bar-Ilan, but I decided to pursue a Ph.D. in neither field. I wanted to make the most of the five years I had studied without becoming a physicist, but at the time I didn't know quite how. I had studied in the framework of the so-called academic reserve (*the Atuda Academit*), which meant that my army service was postponed for the extent of my studies. Realizing I was not interested in a career in physics or mathematics but with no idea of what I did want to do, I enlisted in the regular army for my obligatory three years to serve as an artillery officer. And when my three years were almost up, the Yom Kippur war broke out (1973) and being in somewhat of a key position, I was asked to remain in the army, and I signed on for an additional three years.

So you served as a career officer in the regular army?

Indeed. For six years in all. And it took me all that time to realize not only that I didn't want to become a physicist or mathematician, but that I didn't want to remain a career officer either! So, I decided not to renew my army contract, in order to do two things: One was to study philosophy. Specializing in philosophy of science seemed natural, as I was reading quite a lot of it in my spare time, especially Popper's work. The second thing I wanted very much to do was to become a high school teacher. I chose Blich High School in Ramat Gan, because I wanted to teach in a good secular school—where, funnily enough, I found myself teaching side-by-side with Danny Statman and Avi Sagi, also religious and Zionist young men who were also deeply interested in philosophy. I taught physics and math as well as a special program of my own design in the history and philosophy of science. Danny and Avi taught Talmud. Those were good years. We had a whale of a time with the students, and we enjoyed each others' company immensely.

But back to philosophy.

My philosophical point of departure was thoroughly analytical. I wrote an exceedingly analytical M.A. thesis on the so-called black ravens paradox. But that was soon to change. At the oral defense of my thesis, one of my examiners, historian and philosopher of science, Yehuda Elkana, who later founded the Cohn Institute at Tel Aviv University, asked me outright what I had learned from writing it. "That that was no way to do philosophy of science," I answered. Philosophy of science should not be pursued in the abstract, but in a way intimately connected to science as it is and was lived and done, combined much more closely with the history of science. But how to do so, I admitted, how to glean philosophical "ought"s off historical "is"s, as it were, I didn't have a clue. As I told my examiners, I had just completed a piece of work, of which I was rather proud, while at the same time realizing that that was not the way to go. Elkana loved my answer. He was on the lookout for perplexed young individuals interested in understanding science and willing to think anew. And thus I found myself among the initial invited cadre of young Ph.D. students of the new graduate institute for history and philosophy of science that Yehuda and Amos Funkenstein were founding at Tel Aviv.

I set forth on my Ph.D. recklessly enough. Rather than address the question of how the history and the philosophy of science should relate to each other directly, I decided to do so historically, by looking closely at how some of my philosophical heroes had combined the two in practice. Auguste Comte, William Whewell, Ernst Mach, Pierre Duhem, Emile Meyerson, and Alfred North Whitehead all seriously engaged both the history and philosophy of science, but without ever reflecting philosophically on how their twofold projects were supposed to work. I thought I had a plan for a very neat dissertation comprising chapters devoted to each of the six, which would first summarize their historiographies of science, then outline their philosophies, and finally try to say something about the relationship that in fact obtained between them.

In 1983 I went to Oxford to work on my dissertation with the philosopher L. Jonathan Cohen at Queen's. And it took me but three weeks or so to realize just how crazy a plan it was and that the dissertation would have to be limited to one person only—William Whewell—and that that too would be a daunting undertaking. I wrote a first full draft on Whewell only to realize that by the time I had finished, my initial question about properly relating history and philosophy of science had faded into the background and what had swiftly come to the fore was that of understanding the kind of questions that had prompted *him*. Thus my actual dissertation ended up offering a philosophical interpretation of Whewell's project, in the form of an attempt to reconstruct the deliberative process, to which his highly original finished works owed.

I was determined not to read Whewell's works as the mere products of the cultural forces and intellectual influences to which he was subject, but rather as the products of keen deliberation. But to do so I had to become a historian—mining the archives and sifting through numerous discarded early drafts of his works, with a view to reconstructing the process by which they gradually took form, carefully tracking and trying to explain what was retained, what was rejected, and what was restated from one draft to the next. Whewell himself never reflected in writing on the formation of his thought. There is no Whewell diary or intellectual autobiography. It was up to me to try to reconstruct and articulate the dynamics of his unstated inner dialogue. And the picture I formed in attempting to do so has gradually come to inform my view of philosophy and how I do it.

Your earlier training in the sciences was in physics. Do you think your turn toward the philosophy and history of science would have happened if you went into the biological sciences, for example? It seems to me that your early training in physics and mathematics made it possible for you to explore philosophy of science as you did. But be this as it may, I want to mention another institution that shaped your career: the Shalom Hartman Institute. Given your training in history and philosophy of science at Tel Aviv University and Oxford, what do you bring to the Shalom Hartman Institute?

My association with the Shalom Hartman Institute grew out of a particular set of biographical circumstances. The turning point in this respect was a major conference in honor of my father, held by Bar-Ilan University in November 1989, at which I was asked to present a paper addressing my father's work. I couldn't refuse, of course, but I was at a loss as to what I could contribute. I knew and admired my father's work, which, as you know, focused primarily on English literature and the Bible, but my own teaching and writing at the time, primarily on early Victorian history and philosophy of science, seemed wholly unrelated to anything he had written. But the organizers were insistent. I ended up reflecting on the last of the volume of essays he had just published on the Bible as literature and anti-literature called *Poetry with a Purpose*. The chapter, entitled "Qohelet: A Hebrew Ironist," offers an intriguing reading of Ecclesiastes as an attempt to articulate an ironical study of human rationality when exercised with the hubris of self-sufficiency; as a sort of a Renaissance exercise in which all obligation to a higher external authority is denied and replaced by exclusive reliance on one's own wisdom and rationality. By applying reason to itself, my father argued, Ecclesiastes purports to shows us the utter futility of such an endeavor, exposing it as a vanity of vanities (in Hebrew, *havel havalim*), concluding in its last verse but one: "The end of the matter, all having been heard: fear God, and keep His commandments; for this is the whole man" (12:13). I found his reading both brilliant and wrong. Wrong in the sense that Qohelet could be read not only very differently, but in a way I believed the Talmud (the little Talmud I knew at that time) had read it. My argument rested on the fact that the Mishna's famous mythic account of the controversy concerning the book's biblical status is presented as a disagreement between the House of Hillel, who were in favor, and that of Shammai, who objected. My father's proposed reading of the religious message of the book, seemed to me far more akin to a Shammaite than to a Hillelite approach. I, therefore, reflected at the conference on the possibility of giving the text a different, more Hillelite reading by reading the book's key-term "*hevel*" as connoting the fleeting temporality of mist or vapor, rather than "futility" or "senselessness" as it is normally understood.

Interestingly, the Zohar understands "*hevel*" as you suggested.

I wasn't aware of that. But I'm not surprised, since "*hevel*" can mean "vapor" in biblical Hebrew. When "*hevel*" is read as "vapor," the premise to Qohelet's entire argument is changed: not that all humankind can boast to know and to achieve under the sun is meaningless and futile, but that it is temporal, tentative, and time-bound; here today, but perhaps gone tomorrow.

If Qohelet's unwithdrawn premise is that all human achievement is futile, then the book's main question as to the possible worth of human efforts under the sun is rendered rhetorical. But if the book's premise is that all humanly possible knowledge is forever tentative and time-bound, and hence hypothetical, however certain it might seem, then its leading question as to the possible value of human endeavor under such epistemic conditions, is not rhetorical at all!

How did your father take your criticism?

Well, he sat in the first row absolutely thrilled. It was the only paper that engaged his work critically—in an appreciative and constructive way, of course, but critically nonetheless. Dad was the first person to raise his hand

and ask a question, and my answer to it marked the beginning of a new and moving phase in our ongoing discussion of matters Jewish. It was a lovely moment for both of us.

I can imagine why: because you constructively engaged the critical spirit of both your father's scholarship as well as the Book of Ecclesiastes.

In retrospect, it was a moment that launched me in the direction of thinking seriously about Judaism through the prism of Talmudic reasoning. Reflecting on my father's legacy, one could say that he (not unlike his politically engaged archrival, Prof. Uriel Simon) was a biblical Jew. For it was in the Bible that he sought and found the normative grounding, idiom, and imagery from which he crafted his religious account of Zionism's calling and aspirations. It is fair to say that it was in that paper that I began to find my way as a religiously and politically engaged Jewish philosopher, as well as a Talmudic Jew. The Talmud's inherent dialogism, I found, resonated intriguingly with my Popperian leanings in the philosophy of science. The 1989 paper was expounded a little more boldly in a series of talks delivered at the Van Leer Institute a year or two later, which formed the basis for a little book published in Hebrew in 1995 called *"To Know Wisdom" (Ecc. 1:17): Science, Rationality, and Torah Study*, which in turn formed the basis for *Rational Rabbis: Science and Talmudic Culture*, published by Indiana University three years after that, and in which the initial phase of the project reached fruition. David Hartman's invitation to join the Hartman Institute more or less coincided with the publication of "*To Know Wisdom*."

The philosophical orientation of all those works owed to Karl Popper's philosophy, to which I was firmly committed at the time. I am now working hard on a substantial broadening and rearticulating of the project in the light of my more recent philosophical work that builds on, but especially away from Popper. The first, second, and fourth of the essays collected in this volume chart that development.

Could you say a little more about the mutual bearing of your philosophical and Talmudic work?

Certainly. In a word, the Talmudic literature presents its readers with two perpendicular axes of dialogical engagement, which, I believe, jointly go to the heart of its religious project. The two axes are nicely exemplified in the famous Talmudic story of the halakhic dispute between the Houses of

Hillel and Shammai and its heavenly resolution (BT *Eruvin* 13b). The heavenly voice indeed rules in favor of the Hillelite's approach to halakha, but only after famously declaring that both "these and these are the words of the living God" (*elu va-elu divrei Elohim hayyim*). But if both are deemed viable and authentic voices of Judaism, why is the Hillelite approach privileged? The Talmud's interesting answer is that (unlike the Shammaites) the Hillelites were privileged because they were modest and nondogmatic, and they not only gave the Shammaites their say, but would lend their views precedence in relation to their own—which I take to mean that although they were as seriously committed to their halakhic traditions as were their Shammaite rivals, they were nonetheless forever wary of being wrong; convinced as anyone can be of being right, yet open to the possibility of being proved mistaken. Yet they realized at the same time, and this goes to the heart of my later work, that we are extremely poor self-critics. People are capable of modestly admitting that their normative commitments *may* prove to be wrong, but not of actually *proving them* to be normatively wrong. Normative self-criticism, as opposed to mere self-doubt, requires levels of self-alienation that the Hillelites realized cannot be achieved by talking to oneself, yet can be achieved if you are willing to expose yourself to the normative criticism of others.

In other words, for a Hillelite position to function properly it *needs* serious opposition! If the Talmud wants to endorse the Hillelite's approach to halakhic development, it *must* deem the opposing Shammaite school to be a legitimate voice of rabbinic Judaism and grant it a real place at the table. To paraphrase the story's famous punch line: "*Only if* these and these are the words of the living God, *can* halakha be in accord with the House of Hillel"!

This is an exceedingly interesting version of pluralism. The other isn't merely tolerated, but valued for his otherness.

Exactly, and we shall return shortly to the crucial difference between toleration and pluralism. But let's stay a moment longer with the story. What has been generally overlooked by the famous tale's many commentators is the way it extends the famous dispute from the horizontal axis of religious deliberation to the vertical axis of religious authority. Unlike their namesakes, the two Houses are conceived not as promoting their own halakhic *opinions*, but as bearers and transmitters of Shammaite and Hillelite halakha. Halakhic tradition is not an interlocutor in rabbinic debate, but a vertical source of religious authority. And the two schools differ sorely as to its nature. The Shammaites deem their traditions fixed and irreversible and hence obligated to abide by them and rally in their defense. The Hillelites also deem their traditions compelling, yet as possibly wrong and in need of revision, and therefore deem themselves religiously obligated to holding them in constant review (for which Shammaite critical input is essential), and to revising them when needed.

Viewed thus, the story of the Houses' dispute can be considered a small-scale version of the Talmud's entire enterprise, in which an astonishingly broadly conceived horizontal dialogism, extending far beyond the world of rabbinic disagreement to include even the prostitutes of Rome, can be shown to be pressed into the service of a radically confrontational engagement of all four of Judaism's main vertical sources of religious authority: Scripture, the law, the institutions of law, and the Almighty Himself. My Talmudic work in recent years has focused on systematically teasing out this aspect of the rabbinic literature and rendering it explicit as a theology and form of religiosity quite different from what we have come to expect.

In this reading of Talmudic reasoning, the Talmud becomes a work of philosophical significance that anticipates self-critical modernity.

Indeed. This is where, I believe, my philosophical work engages Judaism at the core of its formative canon.

If so, would you call yourself a Jewish philosopher?

I would, but I would like to reflect a little on the meaning of the term "Jewish philosophy." Narrowly defined, Jewish philosophy, as I understand it, is first of all philosophy. Jewish philosophy is philosophy applied to matters Jewish. But this definition is insufficient. For Jewish philosophy to be Jewish philosophy, it also has to be conducted from within the world of Jewish concern. A good analogy is Michael Walzer's notion of "connected" social criticism. In this model the social critic comes from within the community he criticizes, speaking to compatriots in the first person, about what *we*, not they, should and should not do. By the same token, I think of Jewish philosophy not merely as philosophy done by Jews, or philosophical reflection on matters Jewish *per se*, but as Jewishly connected philosophy: philosophy written from within Judaism and about matters Jewish. For this reason, I consider my philosophical work on and with the Talmudic literature as Jewish philosophy, as a form of philosophical self-grappling with and within the Jewish tradition.

Were you trained as a Jewish philosopher?

No. My initial training was in philosophy of science and in general philosophy.

Even though you did not see yourself initially as a "Jewish philosopher," you do have knowledge of the Jewish philosophical tradition. Right?

Yes, yes.

So, within that tradition, which period or which body of texts do you resonate with the most? Which Jewish philosophical texts appeal to you most?

Among the decidedly Jewish philosophical texts I know, I'd certainly count Qohelet in that regard. Maimonides' *The Guide to the Perplexed* is another important text for me. But there are also other pregnant moments of philosophical interest for me within the Jewish tradition. For example, the Book of Job, which I'd not classify as a work of Jewish philosophy proper, despite its great philosophical significance. And there are several other such biblical moments. Maimonides' halakhic work, while not itself philosophical, virtually bristles with philosophical interest. Take for example, his Laws of Repentance, *Hilkhot T'shuva*. *T'shuva* is a uniquely Jewish mitzvah, requiring one to live in a perpetual state of normative self-examination. It is a religious obligation that goes to the heart of our humanity, namely, our ability to keep ourselves in normative check, and hold ourselves accountable for what we judge to be our normative failings.

What makes Maimonides's treatment of *t'shuva* philosophically fascinating is the awareness it shows of the troubling difference between holding oneself accountable to what one knows is right and questioning one's norms themselves. It is one thing to demand that we impose our norms on our desires and impulses, but quite another to demand that we keep our very norms in normative check, as Maimonides requires.

But how is that possible? How can we be required to normatively question the very norms by which such questioning is performed? Like the Talmud's understanding of the Hillelite approach, Maimonides also clearly senses that such feats of self-alienation are humanly impossible; that we can only ever achieve a truly reflective grip on our own heartfelt commitments when challenged normatively by people differently committed.

For rather than describe *t'shuva* as a painful inner process of private self-grappling, as one might have expected, Maimonides insists repeatedly on *t'shuva*'s public dimensions. Thus he insists that one should articulate one's failings publically, move out of the comfort zone of one's consensual community, change one's name, and most importantly, cease to be defensive and expose oneself willingly to the rebuke of others—all of which bears directly on the main topics of my work.

These are a small sample of Jewish texts I find "philosophically appealing," as you phrased the question. Needless to say, grappling with the opposition—i.e., with the challenge posed by the many less philosophically appealing voices of the tradition—has had a far deeper formative impact on my thinking than that of those with which I find myself in agreement. Finding ways to counter the tradition's dogmatic, triumphalist, intolerant, nationalist, racist, xenophobic, misogynic, and downright irrational voices from within, is what really animates my philosophical efforts in this regard.

Staying with the appealing, what is the next moment in Jewish philosophy after Maimonides where that self-criticism comes to the fore?

As I mentioned above, the philosophical roots and inspiration of my Jewish philosophy come by and large from outside the field—first and foremost from philosophy of science, and later from political philosophy and from the philosophy of mind and language.

If I remember correctly, Popper was Jewish. Right?

Yes, he was, but although he acknowledged being born a Jew, he denied his Jewishness. Thus, he refused to have an entry in his name included in the *Encyclopedia Judaica.* Although his thinking provided the initial key for my work on the Talmudic literature, I don't think there is anything particularly Jewish about Popper's philosophy, which developed as an attempt to make philosophical sense of the great overturning of Newtonian physics at the beginning of the twentieth century. And his great contribution to philosophy is the identification of rationality and criticism, rather than rationality and truth. For Popper, to act rationally is not to confidently attempt to prove one's pet theory, but to humbly assume that it's most probably wrong, and to, therefore, do one's best to test it as rigorously as possible. Science, according to Popper, owes its greatest advances to successful refutations of its findings rather than to attempts to confirm them. This is Popper's great contribution, although he did very little actual philosophical work on the notion of criticism proper: what it is to criticize, how criticism works, what transformative impact criticism can and cannot have, what it is to be criticized, and to endorse one's critic's position. Most importantly, Popper and his school have had virtually nothing to say about the normative dependency of self-criticism, and hence on the troubling limits of self-criticism—the problem that animates the book I published with Yitzhak Benbaji in 2011, *The View from Within: Normativity and the Limits of Self-Criticism.*

And your Jewish philosophy?

My trajectory as a Jewish philosopher was not a simple progression, nor a simple application of my general philosophical work to matters Jewish. It is better described as moving back and forth between the two. Popper's philosophy initially informed my thinking about the nature of dialogue and how the Talmud reflects on it, and it motivated my undertaking of a more comprehensive and systematic philosophical study of the Talmudic literature. This, together with some of the work I was doing in the history of science and mathematics, was responsible, in turn, for the thorough rethinking of my Popperian commitments I engaged in from the late 1990s onwards. And this, in turn, has led to a significant deepening and broadening of my understanding of the Talmud's project and its philosophical significance. And the new understanding of Judaism's rabbinic canon that has emerged has formed the basis for my work in recent years on the religious challenges of sovereignty, my involvement in interreligious studies, and in interreligious dialogue and learning.

The dynamic interplay between my work as a philosopher and a Jewish philosopher is more than symbolized by my weekly shuttling back and forth between Tel Aviv University and the Hartman Institute, Jerusalem, two to three days a week at each for the best of twenty-five years. It has not made me a better Talmudic scholar, nor a better philosopher than my colleagues in either world. By no means. But bringing my changing philosophical perspective, initially honed in thinking about science, to bear on some of Judaism's major sources, has enabled me to propose readings of them rather different than those we've been accustomed to, just as the benefit of my newly acquired Talmudic perspective, funny as it might sound, enabled me to look anew at the history and philosophy of scientific revolutions.

How is your method different from Hermann Cohen's treatment of Jewish sources?

In one sense it is similar to the way Hermann Cohen brings his Marburgian Neo-Kantianism to bear on the Jewish sources. But on the other hand, it is different. Cohen's monumental *Religion of Reason Out of the Sources of Judaism* undertook to reinterpret Jewish ethics and theology along Kantian lines, redefining the idea of God itself, for example, as conjoining Kant's idea in the second critique of a perfect and perfectly universal morality with the regulative idea of its full realization in the end of time. In terms of theology and religiosity, the philosophical perspective my own work purports to bear on the Jewish sources I read has far less to do with their content as with the picture of human reason and humanly possible intellectual achievement attested to by their dialogical and deliberative modes of reasoning and interpreting. My philosophical interest and work in this regard focuses less on the rabbis' first-order rulings on matters halakhic, theological, and moral, and far more on the second-order meta-halakhic, meta-theological, and meta-ethical presuppositions that inform their religious discourses.

But there's a distinctly dialogical aspect to Hermann Cohen's thought too.

There certainly is. Indeed, much of modern Jewish philosophy, beginning long before Hermann Cohen—Moses Mendelssohn and Solomon Maimon are clear examples—is gradually becoming more relevant to my work, not so much for the way it purports to read the Jewish texts of old, as for the definite "dialogical turn" one can detect in the thinking of many of its most prominent voices—among them, in addition to Mendelssohn, Maimon, and Cohen, one can certainly count Buber and Rosenzweig, Heschel and Levinas, not to speak of Soloveitchik, Leibovitch, Hartman, and several others. At present I am collaborating with Christian Wiese, a Frankfurt-based, master historian of modern Jewish philosophy, in an attempt to make sense of this reemergence of Jewish dialogism.

Other historians of modern Jewish philosophy who work on the dialogical philosophers you just mentioned are not working with traditional Jewish texts as you do. Right?

Yes, and understandably so. German modern philosophy is not my specialty; I listen in, and to some extent I'm aware of its major voices and their arguments. But it is not the philosophical tradition in relation to which I matured as a philosopher. The development of my Jewish thinking was idiosyncratic in comparison: arising through the philosophy of science and then returning to the texts of old and working itself up as Jewish philosophy while focused insistently on the Talmudic literature. Given this peculiar path you can understand how and why my approach is different from that of Hermann Cohen and why I am fascinated by the inherent dialogical aspects of the new philosophy that both informed his great work and developed in its wake.

I am a traditional Jew, and the formative canon of traditional Judaism is the Talmudic literature rather than latter-day Jewish thinking. The Talmud is the one work that every learned Jew studied throughout the generations. Reading the rabbinic canon anew, as I think I honestly can, as informed by a view of rationality and normativity, not only intriguingly akin to, but in a manner more knowingly aware of than that of contemporary philosophy of mind, language, and science, is to give it a new life and relevance.

Given this methodology, it seems to me that for you the institutional setting where you teach makes little difference. The educational setting doesn't really change the methodology. Rather, you bring the same methodology to every setting in which you find yourself? Is that correct?

In a sense, that's true, though the didactics might be somewhat different according to the teaching context. Needless to say, teaching a rabbinical text will always be different from teaching a philosophical position or dispute. The former will generally be more text-centered than the latter. But even when teaching a rabbinic text, it's one thing to read it in the light of, or as exemplifying, a philosophical position, and quite another to read it as challenging it. At Tel Aviv University I normally devote one of every four or so of my graduate philosophy seminars to aspects of the Talmudic literature. At the Hartman Institute, my work and teaching focuses on it exclusively.

You are one of the few people who have attempted to deal with that relationship, both systematically and conceptually, and you've done so as an observant Jew. Currently there is a very small group of people who are interested in the conversation about Judaism and science. As you know, Ian Barbour has identified four types of relations between religion and science: conflict, separation, convergence, and dialogue. What do you think about this typology as a historian of science, a philosopher, an interpreter of Western culture, and a religious believer? Which of these types best describes your view on the relations of science and religion?

Well, first, a caveat. Although I love to talk about the relationship between science and religion, it is not a major focal point of my work. The subtitle of *Rational Rabbis* is *Science and Talmudic Culture*, but the book is not about the Talmudic literature's attitude to science or vice versa. It is more about applying insights gleaned off a philosophy of science to understanding the Talmudic literature. My work hardly touches on how the Talmudic literature engages the natural world, and in truth I've written very little on science and religion per se. Indeed, the introduction to philosophy of science I teach at Tel Aviv University sets forth historically from the views of science to which Reformation and Counter-Reformation theology gave rise. Judaism doesn't enter the picture.

In other words, is it correct to say that the discourse on science and religion is decidedly a Christian issue and the perception that there is a problem between religion and science is really a Christian perception? Is it accurate to say that Judaism is more "science friendly"?

"Discourse" is too broad a term in this respect. There's a difference between the problems religion has with science and those that the scientific culture of modernity has with religion. In the first category, I think that, on the average, Judaism, even in its most traditional forms, may fare better than the more traditional forms of Christianity and Islam. In the latter sense, there is no longer much difference. Nonetheless, it is a tension that has gathered momentum in the wake of the scientific revolution of the seventeenth century, in the course of developments in which Judaism (and Islam) played no part. Not that Judaism and Islam would have fared much better if they had, but, as a matter, at least, of historical fact, the origins of modernity's growing impatience with religion and religiosity are to be sought in Christianity's involvement in the rise of modern science, for which, as I explain at length in the third essay collected in this volume, we will do well to consider the work of Peter Harrison seriously.

In a word, inspired by Max Weber's well-known *Protestant Ethic and the Spirit of Capitalism*, most historians and sociologists of the scientific revolution have tended to locate the role played by the Reformation in the rise of science as owing primarily to Calvinist Puritanism. Peter Harrison's groundbreaking *The Bible, Protestantism, and the Rise of Natural Science*, locates it not in Calvin's Puritanism, but in Luther's revolution of Bible study. A major aspect of Luther's campaign against the priestly hierarchy of the Catholic Church was what might be termed a radical "privatization" of revelation, deeming it to be the religious duty of all Christians to read the bible for themselves, and their personal responsibility to receive the word of God directly. The revolutionary idea that the word of God could be read off the biblical text by all and sundry, necessarily required attributing a radical literalism to the text. The word of God, urged Luther, resides in the plain meaning of the words that meets the eye. Peter Harrison's innovation was to show how, by virtue of the well-entrenched two-books analogy, Protestantism's new and revolutionary literalism with respect to the Book of Scripture strongly implied adopting with similar urgency an analogous approach to the Book of Nature, that grounded scientific understanding in the facts that meet the eye, rather in their symbolic or spiritual significance. My own contribution to Harrison's proposal, as detailed in that essay, was to add to it a philosophical twist, by showing how early Protestant biblical literalism yields *empiricism*, a philosophy of science when applied to the book of nature. The paradigmatic "Protestant" philosophy of science, in this respect was, hence, that of Francis Bacon.

Are you arguing that the rise of science was essentially a outgrowth of Protestantism?

No, that's only half the story. What Harrison overlooks completely is the significance for science of the Catholic backlash. God forbid that the meaning of His sacred word be left to the discretion of uneducated common folk however pious and good willing! Maimonides viewed the Bible as the most dangerous source of idolatry when its anthropomorphic depictions of God are read literally. The sacred text, claimed the Catholics in similar fashion, is a matter solely for expertly trained theologians well-versed in the rational theology one must bring *to* the Holy Scriptures and read them in its light. Only when one's theology is in place can the reading of the Good Word begin. The Catholic Counter-Reformation thus insisted on viewing Bible study as a keen process not of reading out (*exe*gesis), but of reading *in* (*eise*gesis), which when applied to the book of nature, again by virtue of the two-books analogy, yields a vision of science that build its understanding of nature, not on the precarious empirical uptake of the senses, but on a bedrock of absolute certainty attested to by reason. Rene Descartes' rationalist philosophy of science thus emerges as the quintessentially Catholic rejoinder to Bacon's empiricist inductivism.

The fundamental disagreement between Bacon and Descartes is hence the philosophical equivalence of the Thirty Years' War, if you wish. Firmly grounded in their respective and diametrically opposed religious cultures, modern science sets forth, inspired by two diametrically opposed philosophies that in ways inconceivable to both, somehow combine and join forces in Isaac Newton's physics, which, in turn, required each of them to rethink its commitments and consequently to inevitably sever its initial religious grounding. Hereby, that essay argues, lies the beginning of the profound decoupling of science from religion in both its Christian formations, which culminated in the conflict between "religion" and "science" witnessed today.

If I hear you correctly, in order to make sense of the relationship between religion and science we need to know the history of science.

As well as those of religion and of the philosophy of science .... And there's also a Jewish side (or non-side) to the story. Years ago, Noah Efron and I wrote a paper about the late-sixteenth-century, Prague-based David Gans, one of the few Jews who were attentive, and even a little involved in the scientific revolution. Gans, a rabbinic Jew with a flair for mathematics, devoted his life to writing books in Hebrew to enlighten Jews with up-to-date knowledge of the science, technology, geography, and history of his day. It takes a while to realize that he viewed the new astronomy (in which he was well-versed), as of all science, very differently from his Christian peers. Christian Europe was obsessed by the question of scientific truth, and hotly enmeshed, after Copernicus, in the great dispute between the advocates of the mutually incompatible geocentric and heliocentric systems of the heavens. For Christian scholars the very aim of science was to discover and secure the one true reading of God's "book of works," just as the very aim of Bible study was for them to arrive at the one God-given truth of His "book of words." Both questions, as epitomized in Gans's work, left Jews quite cold.

Why?

Because their version of the two-books metaphor was very different. Gans's astronomic textbook relates the history of astronomy in well-informed scientific detail describing how Ptolemy the great scientific sage said one thing, after which came Copernicus the great genius who said another, who was followed by the great authority Tycho Brahe, whose opinion was different yet from both, and you suddenly realize that for Gans these theories, though incompatible, were not in conflict with each other, and that he had no interest at all in adjudicating them.

Gans saw himself, we argue, as compiling a midrashic text, as treating the various scientific theories as a rabbinic Jew would study *Tanakh* along with its various conflicting commentaries—those of the Talmudic rabbis of old, alongside such later exegetes as Rashi, Nahmanides, and Abraham ibn Ezra. While each interpreter does his best to determine the text's true meaning, their diverse efforts are collated and keenly studied with a view to appreciating the range of possible meanings to which God's word gives rise, rather than to determine its one true meaning! Analogously, Gans presented the great astronomers' life works, as if they were exegetes, in the traditionally Jewish sense of the term, expounding the book of nature's many possible readings for their students to carefully study and appreciate, but by no means to decide. For Gans the scientist, as for the entire rabbinic tradition of Bible study, it is the polyphonous din of the study hall that serves as the model of intellectual achievement, so very different from the Christian approach.

So what does that mean for Judaism? Should we conclude that the Jewish conversation is not about a quest for truth?

No. Only that Judaism's rabbinic notion of Bible study or scriptural exegesis has since the Talmudic period been problem-driven, yet geared toward the proliferation of possible meanings, rather to articulate and dictate *the* one God-given truth, and that at rare moments in which something akin to the Christian two-books analogy surfaces within rabbinic circles, as in Gans's case, a similar attitude toward the study of nature can be felt.

Are you saying that Jews are more about interpretation and teasing out hermeneutical possibilities that the text provides us, whereas the Christians are more about finding the truth?

Yes, certainly with regard to biblical exegesis—a point brought out powerfully in Maoz Kahana's recent work on eighteenth-century rabbinic reactions to Spinoza's "Protestant" critique of rabbinic exegetical practices. The editors of interpretive tracts, such as the midrashic compilations of old or the *Mikra'ot Gedolot*–style editions of biblical commentary, display the diverse and largely conflicting efforts of individual interpreters, each doing his solemn best to give the text its best possible reading side by side without ever attempting to discern which of them better reflects the one biblical truth. Put differently, in rabbinic biblical exegesis, there are canonical texts, but no canonical *readings*. Inevitably "gappy," like any other text, the sacred text raises interpretive questions that are perceived as opportunities for keen and diverse dialogical engagement, rather than as calling for authoritatively obligating answers. Interpreters engage the exegetical problems they perceive by bringing their own truth to the table and arguing for it, with a view to enriching the conversation, rather than ending it once and for all!

If so, what's the implication for Jews doing science or for Jews engaging in science?

The implications are not for Jews doing or engaging science as such, but for those doing so whose mindset is to an extent religiously informed. In this regard one remarkable latter-day rabbinic text comes to mind—the only text I know in which a major rabbinic authority comes close to articulating the Jewish version of the two-books analogy. It was written by Rabbi Naftali Zvi Yehuda Berlin of Volozhin (known as the NETZiV) during the mid-1860s as part of the introductory essay to his five-tome, verse-by-verse commentary to the Pentateuch, and discussed briefly in my essay on science and religion mentioned above. Addressing his readership of Torah-wise observant fellow-Jews, he writes: "It is easy to understand that just as it is not possible for the wise student of nature ever to boast knowledge of all of nature's secrets . . . and just as there is no guarantee that what his investigations do accomplish will not be invalidated in this generation or the next by colleagues who elect to study the same things differently. . . ." It is surprising to find such a clearly Popperian view of science articulated by a rabbinic scholar who knew very little science, the best of seventy years before Popper. But what makes the text truly remarkable is the way it goes on to speak of Torah study. Rather than contrast the perpetually uncertain realm of scientific conjecture, with the absolute God-given truth of Torah, the NETZiV *compares* the two: "Thus," he continues, just as in science, "it is not possible for the student of the Torah ever to claim that he has attended to each and every point that claims attention, and even that which he does explain—*there is never proof that he has ascertained the truth of the Torah*! Still, we must do what we can." The NETZiV views himself to be under religious obligation to join the discussion of ages, not to justify, but to critically engage his predecessors, even the great exegetes of old, whenever he deems them to have been mistaken. Truth is treated here as a regulative ideal (like absolute justice or beauty) always to be aimed at, though never provable. The only progress we can ever boast in Torah study is by engaging each other as best we can in keen critical dialogue. Just like in science!

In the Orthodox world, who among the disciples of the NETZiV actually followed him? Among Orthodox thinkers who are interested in science the dominant position is quite different; many Orthodox scientists say that there is no conflict between Judaism and science because all truth, including scientific truths, is in the Torah. That means that every scientific development, every scientific insight, or every new scientific theory is already in the Torah. So, how do you deal with these approaches to science within the Orthodox community?

I have little patience with this kind of smug and willingly ignorant apologetic. There are simple folk who believe such empty rhetoric. But I sincerely doubt that anyone half versed in science and rabbinics could be so intellectually dishonest to really believe it.

Do you at least challenge them?

What I say and write about the rabbinic literature certainly does, but I doubt people like that would have heard or read me. Note that to claim that not only everything is in the Torah, but that everything in the Torah is also open for us to see, is to say, therefore, that there is no need at all to study anything but Torah—not only for Torah scholars, but for anyone! This is a view that is not only diametrically opposed to the way I read the Talmudic literature itself, it is a view I find hard to believe that even its most avid advocates would follow *in practice* (as to commission Torah scholars to build their houses and treat their teeth!). As I said, it is the kind of unreflective rhetoric I have no interest in engaging.

In the twenty-first century, our main concern is not just science and relationship to religion, but science and technology. Correct me if I'm wrong, but you have not written a lot about technology or the relevance of technology to Jewish thinking. Why not? And if you were to write about technology or science and technology, which direction would you take?

Hmm, I'm not sure. I've really not given this question much thought. My philosophical work has little special bearing on technology as such. My main interest in philosophy concerns how the normative frameworks that guide inquiry in general can be modified, replaced, endorsed, and adapted rationally. In that regard, although I haven't written about the philosophy of technology per se, or about technology and Judaism, what I think I do have to say about the framework dependency of criticism and the centrality of criticism for all meaningful discourse applies equally well to science, technology, the development and refinement of techniques, of law, of political theory, etc. Indeed, all thoughtful, considered, reasoned action, regardless of realm or discipline, requires or relates to the set of concerns I have with the limits of self-criticism, the nature of normativity, and normative framework transitions.

My work in the history of science and mathematics has focused by and large on nitty-gritty case studies of moments of that kind of intense self-criticism and the framework transitions they yielded. It is all about the rationality of transformative normative rethinking, where human agency asserts itself most strikingly. It guides my work on the Talmudic literature and on the Zionist revolution, but I have so far not applied it to the history of technology or the transformative impact of technology on society as such.

I think that we are in fact in the midst of such a profound revolution right now, because technology is transforming all aspects of human life, including the meaning of being human.

Certainly. And the rethinking of halakha prompted by a variety of technological advances is profound.

Before we leave the discussion about Judaism and science, let me ask you one more question. As you know, the dialogue of science and religion is supported by generous funding from the Templeton Foundation. The two of us actually met at one of the conferences organized by the Templeton Foundation. It is fair to say that very few Jews are involved in these conversations. How do you explain the paucity of Jewish involvement in the dialogue of religion and science?

Again, my answer is personal and somewhat idiosyncratic. We have yet to discuss what I mean by confrontational religiosity, its place in the Talmudic canon, and its relevance today. But to answer your question more directly, let me say that the concerns Christianity has with science are perhaps different from those of Judaism. The Templeton Foundation is a great organization that funds a lot of extremely interesting work. But, as you know, the ultimate goal of the late John Templeton and many of those connected to the Foundation was and remains to explore the bearing of the best of science on the existence of God. The entire endeavor is truth-geared along the lines of the two-books analogy I mentioned earlier.

Judaism, I think, in the spirit of its polysemic midrashic practices of old, has instinctively avoided this pursuit. Even Maimonides' profoundly religious interest in science was not motivated by a desire to provide religious truths with scientific backing or to square science and scripture. There are a small number of Jewish Orthodox scientists who engage in such pursuits, but they are marginal. I cannot think of any Jewish thinker of note whose thinking is seriously motivated and moved by a need to find a role for God in the evolution of the physical and biological world as portrayed by science—a question to which much of the Templeton Foundation's energy is directed. If at all, Jewish thinkers are far more concerned with God's possible role in human history than in natural history—which, in turn, seems of be of little concern among Christian thinkers.

Do you agree with the statement that unless Judaism renders itself scientifically and philosophically, Judaism is about to die, or die out?

Absolutely. All three monotheistic faiths, Judaism included, face a grave challenge posed by modernity to their very notions of religious obligation and commitment. And today Judaism faces two additional challenges that are different, but, to my mind, not unconnected. Modernity's great challenge to religiosity as such has to do with the central norm of disposition shared by all three religions, at least in their God-centered formations, namely, that of submission; a commitment to do one's upmost to discern what one's religion requires of one and to comply with those requirements. And in all three faith traditions, the religious justification for such commitment is one and the same: a firm belief in divine perfection, especially divine moral perfection, that entails an a priori willingness to suppress one's own moral judgment when in conflict with what one takes God's will to be.

Such a willing relinquishing of personal responsibility flies in the face of modernity's ideal of autonomy—a point that was driven home vividly by the late Richard Rorty at the annual meeting of the American Association of Religion in November 2003, who, without mincing his words, declared to the large audience that "people who believe their world is inhabited by an all-good, all-knowing perfect being, whom they believe talks to them, make less better citizens than us folk!" Nietzsche made the same point even more bluntly, and Dewy and Feurbach more politely, perhaps. But the point is clear enough: to bow in submission to a religious authority, not because one autonomously judges its rulings to be good but because one accepts its authority come what may, is perceived, from modernity's perspective, to be at odds with the values of personal autonomy, responsibility, and accountability, and hence as a betrayal of our very humanity.

This is a pressing question for all God-centered forms of religion that cannot be easily pooh-poohed away. Addressing it requires serious philosophical engagement. One way of addressing it is to simply deny God-centered religiosity entirely.

That denial is called secularism, right?

Not necessarily. Religion need not be God-centered, in which case it won't be secularism. It is possible to be committed to a religiously meaningful way of life, in which *religious* law is confined, for instance, to pure ritual. Reform Judaism is a case in point in this regard. Reform Judaism offers a deep and meaningful form of religious commitment, in which God is present but not as lawgiver in the halakhic sense of the word. Reform Judaism took the notion of mitzvah to mean self-rule or self-obligation, yet without rendering itself in the least secularly a-religious. The great challenge of modernity for people like myself—equally committed to a firmly God-centered form of halakhic religiosity as they are to the values of human liberty, autonomy, and moral responsibility—is to determine how to square the two without forgoing either.

Can you reconcile your religious God-centeredness with modernity's notion of autonomy?

I think I can, because of what I said earlier about the horizontal and vertical axes of Talmudic religiosity. The voice of the Talmudic canon my work purports to expose, and to which I am personally committed, equates religious faithfulness and religious obligation with adopting a confrontational, rather than submissive stance toward all of Judaism's sources of religious authority, including God Himself. Such a conception of religious faithfulness is premised, I argue, on an explicit denial of their moral perfection, including divine moral perfection. And, again, it is not merely the claim that our *understanding* of God's morality is imperfect, but the far more radical claim that divine morality *is* imperfect, as is that of scripture, the halakhic tradition, and the halakhic authorities.

In the context of such a *theology of imperfection*, adopting a submissive, obedient stance toward divine command and action and the other sources of one's religion's authority will count as a betrayal of one's religious obligations as a believer! This is so different a notion of religiosity from all current of God-centered forms that most would object to categorizing it as a form of God-centered religiosity at all! The Talmudic canon is by no means a work of systematic theology. But systematic work on it yields, I insist, a bold, and self-conscious form of decidedly non-submissive religiosity that offers a surprisingly real resource for halakhic Jews like myself for addressing this first of modernity's challenges to Judaism.

What do you take to be the two other challenges?

The two other great challenges Judaism faces today are political. As Michael Walzer nicely put it, two latter-day humane solutions have recently been made available to the so-called "Jewish question." One, the full integration of Jews within the Western liberal democracies; the other, modern Jewish statehood. Both represent radically new, non-exilic forms of Jewish existence, in which Jews and Judaism are made truly at home. (I use the term "non-exilic" in the sense that it would be as absurd to describe Jewish U.S. citizens, for instance, as living in exile as it would their Italian or Irish compatriots.) Both new political forms of Jewish existence represent re-entries of the Jews into history as bona fide member communities of equal standing, either *within* democratic nation states or *as* an independent nation state. In both modern political formations, Jews, for the first time in the common era, are charged with full social and political responsibility for society at large, and as such pose grave challenges to our traditional ways of thinking about ourselves and our relationship with, and especially our obligations toward others. Centuries upon centuries of defensive exilic existence, never fully welcome, never enjoying real membership, and, therefore, never held socially or politically responsible for anything beyond their own, have bred a politically and socially blinkered mentality, in which politics for Jews was confined to communal micro-politics, and ethics for Jews, to communal micro-ethics. Perhaps it is not too far-fetched to say that until very recently indeed, Judaism was hiding from humanity, to borrow Martha Nussbaum's use of the term, not entirely unwillingly exiled from history itself, in ways that rendered "good" and "bad" synonymous with good or bad *for the Jews*.

To reclaim our place in history, by partaking again in state policy and politics on an equal footing with others, as sovereigns or as member communities, requires of Jews a radical reorienting of their very normative outlook. In order for the two new forms of Jewish existence to evolve into actual forms of Jewish life in the later Wittgentein's sense of the term—as opposed to treating them as mere comfortable templates for the old forms of communal existence and responsibility—our proactive values (i.e., those that move us to action, not merely to passive, disinterested assessment), have to transcend the communal boundaries of the traditional Jewish "we." This may sound simple, but it represents a major normative reformatting that very many Jews have yet to undergo.

If I understand you correctly, you wish to show how the self-critical nature of the rabbinic canon can offer resources for Jews to engage the challenges of full democratic participation, as well as those of modern statehood. Are you saying that we essentially come equipped to face these challenges of modernity because we already have the tools to do so? Put differently, are you saying that because of them we are capable of actually living up to this unique historical moment?

Well, although it can't get us all the way to sovereignty, as I shall explain, I certainly do believe that, when read as I suggest, what I dubbed earlier the Talmud's Hillelite voice offers a profound resource for the type of pluralism valued by modern liberal democracy. (Although, as I explain in the second essay included in this volume, the arguments for Talmudic pluralism are more Popperian than liberal proper.) However, it is a voice that is now so gravely misunderstood that it has largely fallen off the our religious radar, so to speak. I believe we are capable of facing up religiously to the these revolutionary times, but only if we are willing to re-read our formative canons in significantly new (old) ways.

So you see it as your task to restore that Talmudic moment to Jewish self-understanding.

Yes, but, again such a "restoration" is far easier said than done. It is not difficult to offer such readings to nonobservant Jews on whose radar the texts never figured much to start with. But to radically modify the uptake of a full functioning observant radar is a different story entirely.

So you talk mostly to your religiously observant cohorts.

I talk and write to both—introducing non-Orthodox Jews to the fascinating world of Talmudic Judaism and doing my best to teach those for whom the texts are already formative how to read them anew. The real challenge, lies, of course, with the latter, and it's a real one! But I'm starting to sound as if I thought the political and moral challenges we face are an inherently Orthodox concern. They are not. Democracy poses two problems, not one, and sovereignty, a third, each of which militates against the entrenched ghetto mentality of very many Jews; a mentality that has deep biblical, as well as Talmudic roots, and has little to do with one's level of halakhic observance. In this regard, contemporary Israel is a vivid and troubling case in point. The Bible's master narrative, as it has come to be read, exercises a profound twofold hold on the way Jews have come to envisage statehood and sovereignty—so profound, that it is hard to see how it can be rectified, although, I believe, the Bible's master narrative holds the key to that too.

That sounds paradoxical.

I know. But it's less paradoxical than it might sound. The biblical story of Israel begins with Abraham, to whom the "seed" and the Land were solemnly promised. The people he was to sire were to found not a world religion, but a nation-state, if I be allowed the anachronism for the sake of argument (which, as we shall see, turns out to be less of an anachronism than one would expect). However, although Abraham and Sarah had already settled in the land, the fulfillment of the covenant was promised from the beginning to take an alarmingly circuitous route, leading their offspring first away from the Land to spend four hundred (!) years of bitter enslavement under the heavy hand of Egypt, before returning to it to claim sovereignty. The story told in the five books of Moses is that of the journey of the family of Israel's (i.e., Jacob's) immediate offspring from Canaan to Egypt, where they became, as Pharaoh described them, "*the people of* the children of Israel" (*Am Benei Yisrael*) (Exod. 1:8), from Egypt to Sinai, where in the no-man's land of their desert wandering they became a religion, and finally back to Canaan, where they were to claim it and form a sovereign nation and establish statehood.

You just summarized the Jewish myth but without any quotation marks and without irony.

Yes, I just summarized the biblical story as it is told. Irony will come later. The first thing to notice is that it is the only one of its very many stories that the Bible does not only tell, but instructs each and every Jew to continue telling and retelling, and to solemnly pass on to their children for them to pass on the theirs. And one very special festival night, Passover Eve, or Seder night, is devoted each year exclusively to doing just that. The Bible bristles with foundational and formative narratives—the stories of creation, the lives of the patriarchs, the Sinai covenant, the conquest, stories of judges, kings and prophets, stories of destruction and of salvation—all are important, but none of them comes close to enjoying the sheer constitutive status granted to the Passover narrative. Why? The Bible itself offers the answer, in the way it treats the two parts of the Exodus story: that of the bondage and suffering and that of the miraculous salvation. The former, the painful memory of our cruel enslavement by Egypt, functions as the grounding principle of the Bible's entire, "horizontal" social normative system. The laws of work and rest (Deut. 5:11–13), those of slavery (Deut. 15:12–15), the laws of employment, the principles of personal liability and of equality before the law, the laws of pledges and welfare (Deut. 24:14–22), the law of minority rights and the principles of citizenship (Exod. 22:20–22 and 23:9, and Lev. 19:33–34), and more, are all grounded in what might be loosely seen as the Bible's version of the categorical imperative: do not treat those subject to your jurisdiction as you were treated by the Egyptians when subject to theirs! The Bible's moral code is founded in its entirety on the moral lesson of the Egyptian bondage. In other words, as the Bible tells and uses the story, we were meant to suffer thus under Pharaoh in preparation for assuming responsibility as a sovereign nation. Hence the strange and dark inclusion of centuries of slavery as part and parcel of the very promise of nationhood to Abraham.

The memory of the miraculous salvation from Egypt, the latter part of the story, plays a similar constitutive role in the Bible's master narrative, providing the grounding for the vertical (i.e., man-God) religious axis of the covenant, as exclusively as the memory of the bondage provides for the horizontal. If our commitment and obligations to others are grounded by the Bible in the moral lesson of Egyptian bondage, our religious commitment and obligations to God are grounded in our miraculous delivery from it by His hand. "I am the Lord thy God, who have brought thee out of the land of Egypt, out of the house of bondage. Thou shalt have no other gods beside me" read the powerful opening lines of the Sinai covenant. I am your God who delivered you from slavery, *therefore*, thou shalt have no other gods! And in similar fashion, all other mentions of our obligation to God in the Bible are justified in exactly the same way (Deut. 6:12, 7:17–18, and 8:13–14, to take but a few examples). But it is in the horizontal covenantal dimension that I am interested, for it is there, I believe, that the Bible's original message has been gravely distorted.

In what way?

Well, one particularly clear way of seeing the distortion is to look closely at the biblical ritual of the Omer count—that tense, assiduous daily counting of the fifty days between the first day of Passover and Pentecost. By not assigning a calendric date to the latter, the annual countdown (or count up, if you wish) of the days between the two major festivals ritually drives home the idea—that is by now firmly imprinted on the collective hard disk of Jewish memory—that whatever commenced in the Passover Exodus culminated in what is celebrated on Shavuot. Shavuot of the Bible is the festival of *bikurim*, culminating in the first fruit offering at the Temple. It is a celebration of sovereignty, in which the Israel's farmer citizens travel to their capital in Jerusalem to offer up their first fruits in gratitude. The ritual culminates in the recital of five verses that Deut. 26:3 terms a *haggadah*—five verses that concisely recall the Abrahamic moment, the going down to Egypt, the cruel bondage, the divine salvation, and the arriving and claiming of the Land. And these are the very same verses that the Tannaitic sages of the Mishna would set later as the proof text for the midrashic telling of the story on Seder night! Shavuot, if you wish, is the Bible's equivalence of Independence Day. Celebrating Seder night with Shavuot thus construed squarely in our sites would have lent the story of the suffering its full moral constitutive force. But the Jewish tradition changed all of this drastically.

Sometime in the course of the Second Temple period, Shavuot became *Hag Mattan Torah* commemorating, not sovereignty or statehood, but the reception of Torah at Sinai. The firm connection between Seder night and Shavuot by virtue of the Omer count thus came swiftly and firmly to convey the idea that what began in the Exodus reached completion at Sinai—that Israel was delivered from Egyptian bondage for the sole purpose of receiving Torah at Sinai; or to risk another anachronism, that we were delivered out of Egypt in order to become a religion, in the latter-day sense of the word!

Why is that so problematic?

For two reasons. First, in the Bible's original narrative the goals of becoming a religion at Sinai and a sovereign nation in the Land are not only clearly distinguished as distinct consecutive stages in the making of Israel—the former a midway station en route to the latter—but are intriguingly envisaged by the Bible as delineating different (if nested) social boundaries.

The collective envisaged as standing at Sinai, entering the covenant, and thereby assuming their *religious* identity, are the *Jewish* people: a collective defined by its vertical religious commitments—in short (and anachronistically) a religion. The collective the Bible explicitly envisages as constituting the Israeli nation, so to speak, in relation to which its horizontal, ethical and political axis is defined, is broadly conceived, as comprising other ethnic minorities, in addition to Jews. This is a surprisingly modern conception of Israeli nationhood, which the Bible firmly distinguishes from the Jewish community per se. When the Exodus becomes retold as culminating at Sinai, ethnos and nationhood become conflated by identifying the latter with the former. The Israeli nation becomes the Jewish people, in which the only "strangers" one encounters are converts to Judaism, rather than members of ethnic minorities of the kind the Israelites had been in Egypt. Changing the meaning of Shavuot from a celebration of statehood, or nationhood, to that of "religionhood," to coin a term, thus signals a drastic transformation of the macro-ethics of state law that underlies the Bible's original conception of the horizontal, to a micro-ethics of religious communal life. The Bible's very notions of "thy neighbor" and "thy brother" are thereby radically reinterpreted to exclude all but fellow Jews. The historical story of the actual emergence of the identification of the people of Israel and the community of Jews is exceedingly complicated, and is still being unraveled. Ishay Rosen-Zvi's current work on the genealogy of the category of "goy" is intriguing in this regard. My interest here is not to explain, or even trace, the emergence and formation of the two opposed conceptions, but only to throw them into stark relief and point out the dire consequences of replacing one by the other.

Century after century, era after era of exilic existence, in which Jews have, as a matter of fact, conceived their obligations and dreams of salvation in such purely micro-ethical terms, have left many Jewish citizens of Western democracies, let alone very many Israeli Jews, largely incapable to this very day of seeing themselves as part of mutually responsible "we"s inclusive of more than Jews. This is an incapacity that has little to do with halakhic observance as such.

I see. What is the second reason?

In a nutshell: robbing what I dubbed the Bible's master narrative of its state-political dimension, robs it willy-nilly of its vital *moral* dimension. The story of the enslavement, suffering, and eventual deliverance told and reenacted at each and every Seder has also become the model for the ritual telling of all our many stories of disaster or near-disaster, and actual or longed-for deliverance. In the Bible, the first half of the story, that of our horrendous enslavement, is explicitly remembered to serve as a moral incentive. The heart-wrenching memory of our suffering is meant by the Bible to translate into an angry moral resolve never to treat others as we were treated then; never to succumb to the temptation of becoming Pharaoh-like toward the "strangers" we govern! But when the story is told devoid of its original political context, with Israel becoming a religion at Sinai as its final end, rather than assuming state-political responsibility, the moral lesson of the suffering is lost. The macro-ethics of governing *across* ethnic and religious divides, is replaced, as I said before, with the micro-ethics of communal life, for which the traumatic Egyptian experience is quite irrelevant. We continue to remember the suffering of bondage in great detail at every Seder, of course, but, as I am sure you can personally attest, accompanied by no moral resolve. The ritualized memory of our cruel enslavement no longer translates into ethical conviction. The suffering continues to function centrally in the story, but only as the dark and terrifying backdrop against which God's great salvation is made to shine forth all the more brightly. The greater the suffering, the greater our debt to God, the greater our religious commitment.

And this has become the norm for all memory of Jewish suffering?

Precisely. The way the Seder reenacts and represents the Exodus has become the paradigm for the way Jews conceive and remember collective calamity and celebrate or long for their salvation. The Jewish calendar is dotted with days commemorating a broad array of past sufferings and near sufferings, salvations real and yearned for: the Ninth of Av, Hanukkah, Purim, and now Yom HaShoah (Holocaust Memorial Day), along with several other minor feasts and fasts. Some, like Tishah B'Av, mourn what was lost; some, like Purim, loudly rejoice in being saved in the nick of time. On Passover and Hanukkah, we rejoice more soberly, since deliverance came only after paying an enormous price. Among observant Jews, the normative emotions that these days of remembrance evoke range from grief to thankful relief, but incredibly none are accompanied by even the slightest measure of ethical reflection. This is not to say that they are experienced devoid of commitment. It is only to say that the commitment they do carry is relegated, on the one hand, to self-survival, and on the other, to the vertical axis. Grief and relief, lament and celebration translate into a resolve never again to fall into the hands of a Pharaoh, and into the familiar oscillation between awe and fear of God's judgment, on the one hand, and praise, thanks, and craving for salvation, on the other. The ethical dimensions of Jewish memory are no longer part of the equation.

That's a very worrying thought!

"*Maoz Tzur*" ("Rock of Ages"), the hymn sung on Hanukkah, vividly captures the purely vertical dimension of this manic-depressive seesaw of repeated cycles of disaster and deliverance. The Bible's truly inspiring ethics of memory, to borrow one of Avishai Margalit's potent titles, has been all but obliterated. The empathetic political ideal so central to the Bible's moral system and foundational narrative has largely faded from collective memory, leaving the intercommunal realm of social justice beyond the pale of Jewish religious yearning. This seriously limits our capacity as a people to feel for other victims of such mistreatment.

I can't but see Israel's colossal indifference to the suffering and degradation she inflicts daily on the Palestinian population under her direct and indirect control as owing to this directly. We have become as Pharaoh. My six-year tenure on the Board of B'Tselem, the Israeli information center for human rights in the occupied territories, was a heart-wrenching blur of horror, shame, and moral rage.

This the first major sense in which, I believe, the Jewish tradition militates against both of the new modern forms of Jewish existence. The solution would seem to lie in somehow being able to return to the Bible's original master-narrative. But even if it were possible to reverse the effects of such a deeply set case of collective and ritualized religious amnesia, the biblical narrative itself ends up in diametric opposition to its initial blueprint.

Now you've lost me, I'm afraid. You've been arguing that by turning away from the political dimensions of the Bible's central narrative, it is the rabbinic tradition that is responsible for the ghetto mentality that makes our twofold return to history difficult for so many Jews. So how can the Bible also be blamed in this respect?

Had the Bible stayed with the master narrative with which it set forth in the five books of Moses, you'd be right. But it doesn't. At a crucial point the Bible recoils from its own story, as it were, laying the grounds for the later turning away from sovereignty that is so vividly signaled by the radical redefining of Shavuot.

How come?

Arriving at the Land, as did the Israelites, virtually from nowhere, to establish a state-like political entity of the kind envisaged by the Pentateuch's master narrative, required two major efforts. One is nation building: creating a cohesive national whole out of the different tribes of Israel and the "stranger" minority communities comprising, as the master narrative envisages, the collective that settled in the Land. Cast in modern terms for simplicity's sake, it is the task of creating an Israeli nation, as distinct from the Jewish people, with a shared sense of national belonging and solidarity out of the hegemonic Jewish community and the ethnic minorities. Forging an authentic national "we" is by no means an easy task. Its success, in biblical terms, depended on the ability of the Jewish majority to make the "strangers" among them truly feel at home by viewing them through the morally reflective lens of their former enslavement, to the extent, no less, of loving them like themselves (Lev. 19:38).

The other major effort, of course, is statesmanship. Statehood requires the creation not only of a socially cohesive national body politic, but a stable and viable player in world politics. Statehood requires a Foreign Office, in addition to a Ministry of the Interior, as it were. It has to be capable of entering pacts and agreements and developing commercial and other ties with sister states, for which diplomacy is required. And in order to function within a geopolitical environment it must be capable of defending itself.

The Bible offers us vivid—I would say even paradigmatic—portrayals of each of these undertakings. The Book of Ruth offers an idyllic portrayal of a confident, upright, righteous, moral society, sensitive to the needs of its needy, regardless, if you wish, of race creed or gender; a society in which strangers, even the ultimate stranger, Ruth the Moabite, is treated equally and not shunted aside, say, to allow the Jewish poor the first pick. Moreover, she is made welcome as an equal on the basis of her willingness to join Naomi, and become a member of the society. She doesn't convert in any religious sense of the term (the polytheism of "your God is my God too" is as religiously objectionable as downright idolatry). There is no need for that, and no indication in the text that religion had anything to do with it. In fact, the very idea of conversion is quite foreign to the Hebrew Bible. Indeed, Ruth is described as a Moabite throughout the book without exception. The story of Ruth is a wonderful paradigm of what we might term national (as opposed to religious) inclusion. To the extent that she is made so fully at home that she cannot only be allowed to marry into the Israeli society of Bethlehem in Judah (today many would say that Boaz married out!) but to be proudly portrayed as establishing the royal lineage of David, no less! This is the Bible's most detailed description of its political ideal, not in theory or as a piece of legislation, but as a vivid description or narration of the (imagined, no doubt) reality of Ancient Israel.

With regard to statehood, the parallel biblical paradigm is that of the kingdom and kingship of Solomon. It is truly amazing that according to the biblical narrative, once his throne was secure, King Solomon's very first official act of state was to marry Pharaoh's daughter and bring her to Jerusalem (I Kings 3:1). Until very recently, diplomatic marriage was the standard form for creating pacts and forging treatises. This was Solomon's first priority and a necessary condition for embarking peacefully on his first great state project, building the Temple, which required a complex set of foreign relations and state commerce. Such marriages were not about love or sex, and certainly had nothing to do with idolatry. (The very next verse [3:3] declares that "Solomon loved the Lord!") The temples built for the king's foreign wives on the outskirts of Jerusalem were what we'd today call their countries' foreign embassies, and partaking on occasion in their rituals was no more an act of idolatry than an Israeli dignitary's attending Christmas Mass at Bethlehem is an act of conversion! This was high diplomacy pure and simple, and Solomon excelled in it. (The pact with Egypt, the Bible tells us in I Kings 9:15, paid off extremely well, when Pharaoh captured and ransacked the Canaanite city of Gezer, and presented it to Solomon as a gift to his daughter.) I Kings chapters 3–10 wax poetical in their detailed descriptions of Solomon's Temple, its religiously moving inauguration, and all aspects of state stability and gravitas: extensive commerce, the fleet, the cavalry, the extensive building of Jerusalem, the elaborate administration structure, Queen of Sheba's visit, the treasury, and so forth. And yet, unlike some of Isaiah's prophetic visions, there is nothing imperial about it. Solomon's Israel does not scheme to conquer and colonize its neighbors nor aim at expanding to become a world-power. It seeks to become a thriving state among states, secure and confident, not by the strength of its sword, but the strength of what it has to offer. But then the Bible's tone suddenly changes abruptly, as if suddenly recoiling from the realization of its own very project, as if a different author suddenly took over! For in I Kings chapter 11 Solomon is dramatically re-described as weak and soft hearted, as led astray by his gentile wives, and is bluntly accused of turning his heart wickedly away from God in the very harshest of terms.

The Bible harbors a strong anti-monarchic or at least a critical view of the monarchy from the very beginning as is clear from I Samuel. Does not the critique of King Solomon reflect that anti-monarchic stance?

That's too easy a way out. Between the end of chapter 10 and the beginning of 11 the narrative exhibits a clear and vivid change of heart, not a sudden application of a policy it had had all along. After all, there is nothing anti-monarchical about its depiction of Solomon's Jerusalem up to that point, diplomatic marriages and all! Nor is there anything anti-monarchical in the way the Bible depicts David before him. I prefer to describe it as a recoiling or a radical change of heart. From this point on, the Bible's narrative takes a long plunge to the worse. The kingdom is divided into two separate political entities: that of Israel and that of Judah, both of which are eventually ransacked and exiled.

And after fifty years of Babylonian exile, the Bible offers its alternative happy ending: the return to Zion led by Ezra and Nehemiah and the establishment of a tiny new crown-protected Jewish community in Jerusalem. The so-called "return to Zion" (*Shivat Zion*) reads like a pathetic parody of what the Bible started off with. Ezra and Nehemiah's zealously religious protected Jewish community has no political aspirations to speak of. On the contrary: with its ethnically purist conception of its religious mission, it purges its ranks of all "foreign" women, and refuses to collaborate with the non-Jewish neighboring communities who proposed to partake in the rebuilding of Jerusalem. It is a community wholly invested in the vertical. Its return to Jerusalem is not to create an independent political entity, but to rebuild the city for the sole aim of restoring the Temple cult. It is the story of a reclusive community under the constant threat of its deliberately estranged neighbors, living by the sword, and protected by a temporarily benign Persian regime on which it is wholly and happily dependent. Ezra and Nehemiah's Second Temple is the antithesis of that of Solomon; their return to Zion a parody of the Bible's original master narrative—an Israel in which neither Ruth nor Solomon would have stood a chance! Indeed the last verses of Nehemia closely echo the damning chapter 11 of I Kings.

In other words, although the Bible's original master narrative provides us with surprisingly potent resources for meeting the challenges of full-blooded membership, as well as sovereignty in modern liberal contexts, the Bible's sudden change of heart not only renders them exceedingly problematic, but implies an exceedingly different conception of Jewish statehood that is today swiftly undermining the incredible achievements of political Zionism.

And, again, I am not attempting to describe, let alone explain the shift historically—Christine Hayes's work on how the Ruth model was replaced by that of Ezra is of central importance in this regard—but only to stress the profound discrepancy between the vision for Israel with which the biblical text sets forth and the one it leaves us with by the end of the day.

You present the so-called "return to Zion" as a pathetic decline from the culmination of the Bible's original political narrative in the First Temple, but there are many Jews today who look at the situation of the returnees to Zion as the true political understanding of Judaism. Whereas you are saying that the new community in Jerusalem was an unfortunate deformation of the initial plan, some people say that this is what Judaism is all about, namely, worship of God in the Holy Land through allegiance to the sacred text.

I agree. Indeed, it is not merely a matter of how one reads the Bible, but about the sociopolitical archetypes that have come to inform our political thinking and yearning. Zionism, though brought about by secular Jewish rejection of traditional Judaism, was welcomed by very many Jews as the long-awaited answer, or the beginning of an answer to their prayers. But, and this is the great "but" that must be realized by anyone for whom the prospects of a truly democratic Israel are valuable, what Jews have been praying for centuries, is *Shivat Zion*, not the Israel of Solomon and Ruth. Ezra and Nehemia's return to Zion has become the paradigm of the Jewish renewal of politics, despite the great discrepancy between the Bible's own accounts of the Second Temple's social and political ideals and those of Ruth's Bethlehem and Solomon's Jerusalem. One can see this clearly in the culminating prayer of the section of *shofarot* of the Rosh HaShanah liturgy, in which Jews have and still pray fervently for God to "sound the great shofar for our liberty" and raise the flag of our ingathering from all our exiles, and to bring us to Jerusalem, the site of our Temple, not in order to establish statehood, nor to renew the kingdom, but for the sole purpose of enabling us to perform our vertical sacrificial duties!

Are you critical of it?

Criticizing history is futile. We can't change what the Bible ends up with, but it's of crucial importance, I think, to appreciate the challenge of the Zionist revolution by understanding that rather than being the answer to our prayers, political Zionism is at serious odds with what Jews have prayed and craved for almost two millennia—the fact that political Zionism coincides beautifully with the Bible's original master narrative notwithstanding!

Biblically inspired, even if not directly motivated by what I have termed the Bible's master narrative, modern Israel envisaged and eventually established itself much in line with it: as a state whose citizenry extends well beyond the boundaries of its Jewish constituency, with Solomonic aspirations to become a bona fide player in world politics commercially, diplomatically, and culturally. But nourished by centuries of exilic Jewish existence beyond the pale of political normalcy, the antithetical darker biblical myth of *Shivat Zion* lived on deep within the DNA of the Israeli Jewish psych, nourished by the devastating traumas of the Holocaust and casting an ever-darker shadow of a defensive fortress mentality over political Zionism's initial plan. The profound exilic mentality of *Shivat Zion* remains etched so firmly in the collective memory of the Jewish people as the ideal of Jewish existence, that, as Michael Walzer's recent *The Paradox of Liberation* argues, much of what Zionism has achieved is already being undone. This has little to do with religion per se. Although political Zionism was essentially and self-consciously a secular undertaking, and many of its visionaries and promoters did their utmost to disassociate themselves from rabbinic Judaism, they sought Jewish authenticity by returning to the Bible, largely unaware of how the Bible narrative mitigates against itself in this regard.

If I hear you correctly you are critical of this vision. This is not the ideal of Jewish statehood that should be allowed to shape Israel's dreams in the image of the small, beleaguered community in Jerusalem after the return from Babylonia.

Exactly. So if we're talking about the challenges of modernity and the work Jewish philosophy needs to do to respond to those challenges, I would say that as far as the Zionist reclaiming of sovereignty goes, it is massive. To respond to that challenge we need to go back and reread the Bible as I reread the Talmud: to rediscover the rich resources of the Bible's master narrative, and enable the shining forgotten paradigms of Solomon's Jerusalem and Ruth's Bethlehem—rather than the tiny, exilic, chauvinistic Jewish Jerusalem of Ezra and Nehemiah—to shape our present. But to read the biblical text thus, not as an academic hermeneutical exercise, but as a normative resource, as offering, nay as enforcing a religious incentive, is to read it religiously against itself, against its lived and living meaning. Ruth has been read since time remembered as a story of conversion and religious faithfulness, wholly stripped of political, nation-building significance, while Solomon's Jerusalem is read, already by the Bible itself, as responsible for setting in motion the long sinful decline that culminated in destruction and exile and the eventual Return to Zion led by Ezra and Nehemiah, as its salvific happy end. For Jews today to read the text anew canonically as I propose, is to read it, as it were, against itself.

I understand how the Bible functions in this project of reconstitution of Jewish political existence through selective reaffirmation and reinterpretation. But what is the role of philosophy in this project? Do you read the biblical text as a philosophical text? Is the Bible a book of philosophy as some people like Yoram Hazony, for example, say?

No, not in the sense that Yoram Hazony claims the Bible is a philosophical text. My view is closer to the approach of Michael Walzer and Menachem Lorberbaum, who reconstruct the Jewish political tradition by rereading the biblical text (among many others) through a political philosopher's lens.

One way to do so is to ask how the Bible envisages the ideal society. And what swiftly becomes apparent, in a manner so wonderfully characteristic of the biblical narrative, is that the picture that emerges is troublingly ambivalent, and, I believe, deliberately so. For one thing, its narrative chapters by no means relate a story of success. They certainly don't have a happy ending. On the contrary: the biblical story of Israel's foundation and formation is that of a colossal failure; a largely unsuccessful attempt to establish a religion, as well as a wholly failed attempt to establish statehood! As a religious canon the biblical narrative represents, if anything, a fervent call for self-reckoning. The way its grand master narrative ends up collapsing into the tiny state-protected community of Babylonian returnees to Jerusalem for the sole purpose of vertical worship, is simply amazing.

So, the biblical project ends in a failure but then comes rabbinic Judaism.

Yes, but first a firm caveat regarding the "then" in your question. Again, I do not mean to imply (God forbid) that, historically speaking, rabbinic Judaism began where biblical Judaism ended, or anything to that effect. My objective is first to sharply distinguish and contrast three very different models of Jewish self-understanding presented by Judaism's traditional canon of old that I find highly relevant, for better or worse, to the challenges we face today. It is only in this sense that rabbinic Judaism's model "comes after" the other two.

Rabbinic Judaism indeed presents a very different story from that of the Bible's narrative, which, as I noted, has much to offer with respect to the challenges of modernity, though not in the realm of state politics. If the two new forms of Jewish political life—full membership in the Western democracies and independent statehood—represent dramatic Jewish reentries into history, rabbinic Judaism signifies Judaism's firm decision, in the wake of the religiously devastating bloodbaths of the second destruction and the Bar Kochba revolt, to resign resolutely from state politics and exist and flourish as best it could outside of history. To rise to the challenge of reentering history, we really have no choice but to return to the Bible, acknowledge the failures with which it presents us, and to see them—just as the rabbis of late antiquity did—as our religious obligation to correct.

In his papers from the late 1940s and early 1950s, before despairing of the Israeli rabbinate rising to the historical occasion, and of the State of Israel to its moral obligations, Yeshayahu Leibowitz waxed poetical about the religious significance of Zionism. He especially stressed our great privilege in having to rewrite whole tractates of the Talmud. Now, with real statehood in the making, engaging in real state commerce and diplomacy, establishing a real state legal system, and so forth, all in real historical time (*ba-zeman ha-zeh*, as he put it), the Talmud's idealized unreal and wholly impractical halakhic envisioning of the laws of kings, those of peace, war, and army, those of priests and sages, of private and public law, as well as those of the legal system itself, had to be all thoroughly rethought and rearticulated. This is very much my position too. The Talmud, in teaching us by extended example that a generation's obligation toward the Torah it inherits is to rethink it thoroughly in the light of its own circumstances and perspective, charges us with the religious obligation to rethink and rearticulate it, just as Leibowitz insisted. And a fitting point of departure, to which I believe he too would have then agreed, although he largely missed, is Solomon and Ruth.

And the model to which we should return is the Solomonic model?

I believe we should return to Ruth and Solomon, not as models, but as powerful points of departure. It would be silly to claim that texts dating from the sixth century B.C.E. could anticipate twenty-first-century notions of statehood and nationhood. But they do contain extremely inspiring early visions of a just and decent society, in the case of Ruth, and of flourishing statehood, in the case of Solomon. Of course, our latter-day notions of both are very different. There is no biblical parallel to the emergence of modern individualistic concepts of citizenship and the ideas of civil and universal human rights. But again, if you read the Hebrew Scriptures religiously as the story of the failure of a covenantal mission, you will be reading it very differently than by harking back to *Shivat Zion* as a political ideal for the

Jewish people, rather than as the tragic ending of the biblical chapter of its story. That's the philosophical sense in which I read the Bible.

I'm not sure it is correct to claim that the Jews had stepped out of history after the fall of the Second Temple. As a historian, I find this formulation to be very problematic and I do not think we have to accept it. Jews remained in history in the rabbinic and post-rabbinic periods and have continued to respond to historical challenges although they had no political sovereignty.

Of course. The Jews certainly continued to function historically during and after the Talmudic era. What I meant was that after the destruction of the Second Temple, the emergent rabbinic Judaism radically redefined Judaism's this-worldly, covenantal mission as no longer geared toward independent statehood. The idea of Jewish state rule (and with it that of the restoration of the Temple cult) was relegated to the status of a messianic dream, to after the end of history, so to speak.

At the time Christianity was adopted as Rome's state religion, rabbinic Judaism was radically redefining Jewish this-worldly aspirations as relegated to communal existence that was seldom very firmly tethered geographically. Jews went where the winds of history blew them. The Jerusalem-based priestly cult-bound Judaism of the Second Temple was dramatically refashioned as a religion of prayer and study practiced in easily relocatable locales. Zion is where the Torah resides, states one early rabbinic text with reference to the newly established fifth-century Babylonian centers of study. This was a revolutionary development that ushered in a new phase in Jewish religious life. While largely irrelevant to the challenges of modern statehood, I see the Talmudic revolution as offering a new epistemology of humanly possible achievement, and with it an account of human rationality and normativity which, if I read it correctly, is surprisingly sober and surprisingly modern.

In what ways?

The modernity of the Talmud is found in its acknowledgment and even promotion of normative and ethical diversity, and the discursive framework it provides for taking pluralism and diversity seriously. The Talmud offers us a theology and religious anthropology that views pluralism and diversity not as a problem, but as a religiously meaningful resource, and notions of religious obligation and faithfulness, radically different from all other religious canons. All of this, I believe, are crucially important resources for all the three challenges of modernity of which I have spoken.

Let's explore that point further.

In a nutshell, the most pertinent challenges with which the Jewish reentry into history must deal are those of somehow transforming the micro-ethics and micro-politics of exilic Jewish communal existence into the kind of macro-ethics and macro-politics required by the two new non-exilic forms of Jewish existence. For many centuries Judaism has stood to the world as a sort of alter-ego, or, perhaps better, counterculture, presenting an alternative, almost antipolitical vision of peoplehood that pertained to political formation, to theology, and even to science. Paradoxically, perhaps, the Talmudic literature, I believe, contains valuable resources for reentering history in the manner I described, despite the fact that it was the most pertinent product of Judaism's knowingly exilic turn, and the very linchpin of exilic Judaism's formative canon!

I understand why you would see Judaism as "counterculture" to Christianity, and I can somewhat see it (although to a lesser degree) applicable to Islam. But can Judaism be said to be a counterculture to Buddhism or to Hinduism?

I'm not a historian of religions, but I still think that during its many centuries of exilic, communal existence, Judaism could be viewed as a counterculture to all world religions, if only for its mentality of chosenness and total lack of religious universalist aspiration. Unlike Christians and Muslims, Jews never felt religiously obligated to spreading the Good Word, as they understood it, and harbored no this-worldly yearnings for a world converted to Judaism. This applies equally to the voices within the tradition that interpret the giving of Torah in the no-man's land of the wilderness as attesting to its universality, deeming Torah not as given exclusively to Israel, but to all humanity (*le-kol ba'ei olam*), and the gentiles as liable to severe divine punishment for not living up to it. Here and there sources taking this line go on to welcome studying Torah with gentiles, but I know of no such source that deems proclaiming the Good Word to the world as a matter of religious obligation, or converting gentiles to Judaism, a religious mission.

What the Talmudic canon does convey, I believe, is an intense religiously motivated interest in engaging the civilized world dialogically—not in order to spread the Good Word, and to teach the world, so much as to learn from it! Here, I believe, lies an important and religiously significant key for transforming traditional Jewish micro-ethics and micro-politics into macro-ethics and macro-politics. Unfortunately, the temptations of retaining an exilic mentality are enormous, even here in Israel!

That's why your work on the rationality of Talmudic discourse is so important, because you show that self-critique and self-examination are built into rabbinic religiosity. In rabbinic Judaism the rational and the critical overlap; one has to be both rational and critical in order to fathom the divine will.

Yes.

Is this insight your main contribution to the our understanding of rabbinic Judaism?

Yes, but it needs to be taken one crucial step further. To truly appreciate the rationality of Talmudic discourse one must realize that the Talmud was also well aware (as Karl Popper, for instance, never was), that you cannot be properly self-critical merely by talking to yourself. Truly normative self-criticism requires opening oneself to the criticism of people whose norms differ significantly from one's own. Hence, it is to realize that one's religious obligation is not to defend the truths of one's faith in the face of such external critique, but, on the contrary, to press that critique in the service of putting one's faith to the test!

Instead of the comfort and confidence of God-given truth, Talmudic religiosity commits believers to the restlessness of perpetual self doubt. The sort of self-critical uneasiness that motivates the Talmud's dialogism, as I read it, differs categorically from incentives for dialogue that derive from a religious obligation to proclaim and spread the word. While most religious people worry constantly about getting the Good Word right, Talmudic religious self-doubt, as I put it in the fourth of the essays collected here, extends to the Good Word's very rightness! What is denied by its major voice is the presumed perfection of all sources of Judaism's religious authority. So that looking back to early rabbinic Judaism it emerges not only as a religiously motivated stepping out of history, but as a form of religiosity that diverges so radically from almost everything we now take religiosity to stand for that most people, including most observant Jews, would fail to recognize it as a form of religiosity at all. But if I read the Talmud correctly, it represents a form of rational undertaking more self-consciously self-critical than even science!

What do you do, then, with the nonrational, noncritical, and highly mythical dimensions of Judaism?

I engage them best I can. But the interesting question is not what I do with them, but what the Talmudic literature itself does with them. As I noted above, the rabbis were well aware that a properly Hillelite position cannot flourish by deliberating solely among its own. A robust Hillelite religiosity requires opposition, for it is based on the constant challenge of diverse opinions different from its own, and therefore welcomes a plurality of contrasting voices that are granted a real place at the table—prominent among which are voices of the kind you describe. A position need not be rational and self-critical to pose a fruitful first-order challenge to positions that are. The second-order dispute between self-critical and dogmatic approaches to Judaism may not be resolvable, but the value of the opposition to the former is independent of any such agreement.

On the other hand, religious dogmatism from both within and without has had a corroding effect on Talmudic religiosity. As Boyarin forcefully points out, as the Babylonian Talmud was taking form as an open-ended, essentially undecided, diverse plurality of voices, Christianity chose to found itself, from the Council of Nicaea in 325 onward, on an obligatory set of dogmatic principles of faith. What is important to realize is not only the diametric difference in this respect between the confrontational form of religiosity advocated by the Talmudic literature and the submissive religiosity preached by other Jewish voices, post-Nicaean Christianity, and later Islam, but the strength and depth of the temptation of the latter. The temptation of a religious system one is not only not required to question but is compelled to view as morally flawless and to submit to uncritically is enormous! It is little wonder the Talmud's submissive voice ended up winning the day in the long run.

But Judaism too, starting with Maimonides, defined its list of dogmas that all Jews must affirm, even though after Maimonides Jews continued to debate the precise list of required dogmas. So what do you do with the dogmatic dimension? What do you tell Jews who insist that if one is Jewish one must affirm certain beliefs?

Maimonides is not the best example in this respect, because he at least was well aware of the difference between the truths he believed in and those he affirmed because they are necessary for the less philosophically minded majority of believers (which he dubbed, somewhat misleadingly, "necessary truths"—*amitot hekhrehiyot*—meaning necessary for communal stability, not necessarily true). It remains an open question among Maimonidean scholars which of his famous thirteen principles he actually believed were true.

But even if we bracket Maimonides, as I said before, by the early middle ages, Judaism had indeed abandoned its Talmudic roots and had come far more to resemble Christianity and Islam in its submissive belief structure, and presuppositions of divine perfection. My attempt to reintroduce Talmudic religiosity as a religiously viable option rests on the fact that, although (or perhaps, because) it is largely misread, the Talmud remains rabbinic Judaism's formative religious text. In highlighting its self-critical and dialogical spirit, I'm certainly going against what Judaism has become, but I do so by harking back to what I believe it used to be, by purporting to offer an honest and viable reading of the Talmudic text. But it is not at all easy, as some of the extremely learned defensive attacks on my *Rational Rabbis* amply prove.

You may be right about the philosophical and rational dimension of rabbinic religiosity, yet your interpretation of rabbinic Judaism will be wrong if it does not take into consideration the nonphilosophical dimension that came into full expression in Kabbalah.

I do take those voices very seriously. But I'm not sure what you mean by taking them "into consideration." The Talmudic position to which I am trying to give new life has long fallen off the radar, as it were, and has been replaced by a broad range of "Judaisms," some halakhic, some not, some mystical, some not. They almost all grudgingly accept each other, at least de facto, but none *value* such a plurality (and more) as a matter of religious principle, as does Talmudic Judaism. Nor do any of them boast theologies capable of grounding religiously the two new political formations of contemporary Jewish life, as does the Talmud. So yes, I certainly take them seriously, especially the challenges they pose, while advocating Talmudic "Hillelism" as best I can.

What about foundationalism? If I understand your philosophy of science correctly, what you take from Popper is primarily his anti-foundationalist position, which you find also in the Talmud. The Talmud is anti-foundationalist because it is nondogmatic, self-critical, and inherently dialogical.

Yes, absolutely.

Now, that's a message that most Jews today would either not understand or wouldn't like to hear. And so, how do you make this anti-foundationalism something that Jews could really resonate with?

First, by making a rigorous text-based case for it being there and then by showing its latter-day relevance and value. The first is to make a descriptive, interpretive claim as to what I believe the texts in fact say, the second, to make a normative judgment regarding their possible latter day significance. However, we need to be more cautious when using the term "anti-foundationalism," and here is where I part company most conspicuously with Popper. But to explain my position I need to say a little more about criticism and self-criticism.

Please, go ahead, since it would lead us to talk about textual reasoning and Scriptural reasoning which is the next set of questions.

Let's start with rationality. There are two major schools of thought concerning rationality. The currently most popular, largely game-theoretical approach, takes its cue from economics and equates the rational with the beneficial. Action is considered rational insofar as it maximizes the actor's utility or, to put it less technically, insofar as he would deem it worthwhile. I resent equating the rational with the profitable. Nor do I think that rationality should be a category merely of successful action. I belong to the school of thought that deems an act rational if it is done for a reason. To act rationally, on such a showing, is to take action because one deems it appropriate. Therefore, to act impulsively, instinctively, or unthinkingly, however beneficial the outcome, is to act at best a-rationally. To act for a reason, is to intervene in one's world or one's picture of it because one feels that something about them is sufficiently amiss, or in the danger of becoming sufficiently amiss to require one's intervention. Hence the inherent connection between criticism and rationality. To act rationally in the sense of acting for a reason is to take critical stock of one's situation with a view to do one's best to putting right what one deems to be sufficiently lacking.

But to act rationally involves more than just taking a critical view of things and acting accordingly. For while things might well be seriously amiss, it might be totally inappropriate for me to be the person to intervene. To act rationally requires taking critical stock not only of the state of affairs in question, but of one's ability, capacity, and qualifications to take the required action. Furthermore, a truly rational agent is not only wary of being the right person to take the required action but is also wary of her very judgment. Just as things may go wrong, and may require us or others to put right, rationality demands that we treat ourselves as liable of being wrong. Rationality requires that we do not take things for granted, neither without nor within.

So it's always self-reflective.

Yes indeed. It's always self-reflective. But what interests me in particular are the inherent normative dimensions of such reflection. For to criticize, that is to find things in some way amiss, lacking, wanting, or wrong, appropriate standards of impropriety are required. To doubt is to merely wonder whether something might be wrong, but to criticize is to attempt to *prove* it wrong, and that can only be done against the backdrop of a norm or standard of wrongness. The kind of critical reasoning capable of prompting one to act cannot be hypothetical. Criticism resembles existence proofs in mathematics: "Because A, B, and C," a critic will argue, "therefore, Houston we have a problem!" And to attempt to prove wrongness, a standard of wrongness cannot be merely assumed, it must be solidly in place. To criticize requires taking certain things true and is, therefore, always piecemeal. And because to criticize is to pass *normative* judgment, it is necessarily framework-dependent. One can only properly and effectively criticize, that is to say, attempt to move oneself or others rationally to action, on the basis of the value system of norms of propriety to which those criticized are firmly committed. In this important sense, criticism, and with it our capacity to act rationally, *cannot* be considered a foundation-less enterprise. Moreover, criticism, and hence rationality, in being inherently and inescapably framework-dependent, are also framework-relative. Normative systems vary and differ in all realms of human life. What an Orthodox Jew might find scandalously wrong, a liberal Jew might find admirable, what in one scientific era would be regarded as scientifically preposterous (the idea of relativity of time, for instance, or a moving earth), are considered perfectly appropriate in another.

But what if my norm is a dogma, for example, the belief that God revealed himself to Israel at Sinai?

Your norms are always dogmatic in a sense, precisely because you're committed to them! The norms that are ours are not something we merely *know* or *assume*. Normative commitment is not entirely voluntary, as accepting or rejecting an hypothesis. The norms we're committed to are the norms to which we hold ourselves and others accountable. They define who we are by defining who we aspire to be. Therefore, they belong among the taken-for-granted truths that define our world, in the category of what Wittgenstein dubbed "framework propositions": that is, the truths and principles constitutive of the thinking and judging of those whose framework it is. Our value systems do not remain constant, of course, and our normative commitments are liable to change from time to time, at times dramatically. But at any given moment, the standards by which we judge ourselves and others, are, in the act of judging, held unflinchingly—dogmatically, if you wish.

Which raises the thorny question of the rationality of normative commitment, and especially of that of normative change. For to endorse, modify, or reject a normative framework rationally requires taking critical stock of it and finding it either normatively fitting or normatively wanting. But how is it possible to criticize the very norms that serve one in so criticizing? How, as I put it in the fourth essay collected in this volume, can a normative framework be impeached by its own lights?

Could it be that instead of reflecting on the basis of a norm or norms, what you have is something more basic such as a love relationship? In the case of Judaism we talk about Israel's love to God. That love is not a rational decision, but rather some kind of commitment, a foundational commitment that forces you then to do everything else.

Yes, that is true. It is also well put, as you chose to speak of love rather than fear of God. Acting out of fear, or for the promise of reward, etc., is different from being committed to do the right thing because it is right. The anchoring of normative commitment in caring, and ultimately in love, forms the cornerstone of Harry Frankfurter's philosophy of personhood and agency, to which I largely subscribe. We love and care for the things we conceive as important to us. Certain beloved people, ideas, and objects take precedence over others, and commitment to them governs and regulates our other commitments. But even those "hypergoods," as Charles Taylor calls them, our scientific, ethical, religious, and political master norms, are also modified and replaced from time to time. And we would like very much to think that when modified or replaced it is done for a reason; we would very much like to believe that we not only do what we believe is right, but that what we believe is right is indeed right! But no one, it seems, can ever achieve the level of self-distance or self-alienation needed to normatively criticize one's basic commitments. Or cast in Harry Frankfurt's terminology, we are incapable of willfully changing our will, or of willfully falling in or out of love.

To the best of my knowledge, all philosophers I know who take framework dependency seriously consider the problem of rational framework transitions insurmountable. Thomas Kuhn, for example, chose notoriously to liken scientific paradigm shifts to acts of religious conversion, or gestalt switches, while Wittgenstein, in his posthumous *On Certainty* likens them to the transformative effect that school teachers have on young children, and missionaries on natives! Wittgenstein and Kuhn simply couldn't imagine the possibility of critical reflection on one's norms that is not circular, because those norms are what one uses in order to critically reflect in the first place. It is to this large question that Yitzhak Benbaji and I devoted *The View from Within: Normativity and the Limits of Self Criticism* (2011). We concede that it is indeed impossible for a person to adopt a normatively critical stance toward his very norms on his own accord. However, the normative critique of trusted others, though powerless to convince, we argue, is nonetheless capable of sufficiently destabilizing the norms questioned for us to consider them critically.

If I hear you correctly, you offer a positive view of skepticism from within a religious outlook. Put differently, you religiously affirm the importance of the skeptical. Right?

Yes. It is a view I have done my best to articulate and develop in all generality, but one that in the realm of religion I believe the Talmudic literature implicitly anticipates and forcefully advocates. My own work in this respect is not about religious norms in particular, but about normativity as such, and is hence equally relevant to science, politics, and ethics, as well as to religion. In all of these domains and many more, the decisions we take and the actions we perform are governed by standards and norms, many of which change, even drastically, over time. A quick look through the family album will prove to you how radically our aesthetic judgments are liable to change, and just how surreptitious such changes are, much in line with how Kuhn and Wittgenstein characterize framework transitions. But then changes in aesthetic taste are not expected to be introduced for a(n aesthetic) reason. But politics, ethics, science, and religion, we would like to think, are different. In fact we *know* they're different. They must be different! The thought that political realignment, religious changes of heart, and scientific revolutions cannot be considered rational (i.e., undertaken for political, religious, or scientific reasons) sounds preposterous! Exposure to the normative criticism of others, Benbaji and I argue, is very much like listening to yourself on a recording. For the way they argue against you allows you a glimpse of how they portray your commitments, just as listening to a recording of your voice allows you to experience how you sound to others, which can be very different from your own self-image. The full philosophical analysis we offer of the transformative potential of external normative criticism is slightly more complicated than that, but that's the gist of our argument.

Can we educate people to be more amenable to that form of self-criticism?

I think so. The rabbis certainly thought so, for this is the master idea that drives the Talmud's entire project. Halakhic, theological, and exegetical truth is pursued in the Talmudic literature not by intense self-deliberation, but by keenly engaging one's tradition (the halakhic, theological, and exegetical legacy of the *Tannaim*) by keenly engaging others. Interestingly, as Ishay Rosen-Zvi has recently shown, the rabbis viewed the horizontal reach of religiously meaningful dialogical engagement not only as extending out to the far reaches of the pagan world, but all the way in to grappling with one's "evil inclination"—which in rabbinic anthropology takes the unique form of a halakha-wise trickster with whom one is required constantly to contend. Unlike Greek accounts of the self, the rabbis located the source of evil not in one's base bodily desires and impulses, so much as in the insistent reasoning of a dark alter-ego figure well-versed in halakha determined to lead one astray by cleverly pointing out enticing halakhic loopholes.

What is interesting about this internalization of Talmudic dialogism is that although the *yetzer hara* is considered the source of all evil, unlike the animal passions of Greek anthropology, which we would certainly be better off without, the distinct impression is that for the rabbis, the constant challenge of disputing one's *yetzer*'s reasoned subversions makes one a better person and a better Jew! I doubt the Bavli's fierce dialogism can be seen as a merely stylistic device. It is very difficult to read the text as *not* having been intentionally fashioned by its redactors to educate its readerships along the lines I have indicated.

Until very recently, Talmudic Judaism was the only discursive culture of old, religious, or other that knowingly considered systematic exposure to the normative criticism of others (within and without) a necessary condition for its inner flourishing and enrichment. Nonetheless, from their very inception, all three monotheistic faith traditions interacted with each other constantly—reacting to the (mostly distorted) images they formed of each other and mutually informed by the mostly unconscious transfer of hermeneutical, theological, liturgical, and legal content across interreligious divides, by virtue of effective go-betweens and/or shared mediating languages.

As a historian, I find this observation very interesting and I think it is borne out by what we know about interreligious encounters, especially in the Middle Ages.

It is and more. The three religions engaged each other at all times and on all levels of religious life, to the extent that it is impossible to properly understand any significant moment in the life of any one of them without taking into serious account the subtle give and take the community under study was conducting with the other two, much of the time unknowingly. This requires that they be studied under the same roof by experts of each working in collaborative dialogue with one another. And in view of its complexity, it is a project no university can properly undertake alone.

This is the idea that motivates the Center for Religious and Interreligious Studies (CRIS) that my colleagues and I are working hard to establish at Tel Aviv University in collaboration with like-minded colleagues at the universities of Cambridge and Frankfurt. CRIS will comprise a major research institute dedicated to the interreligious study of the three Abrahamic religions. Over the last three years we've held a series of international workshops and conferences, some at Tel Aviv University, some abroad, and a community of senior researchers and graduate students here and there is swiftly taking form. We hope to be able to establish in the near future a new journal as well as a monograph series in interreligious studies in cooperation with our two partners, as well as properly funded collaborative research teams.

CRIS will also boast a teaching center where a new M.A. program in interreligious studies will take center stage, and in which young Jewish, Christian, and Muslim Israelis will study the formative role played by the two other religions in the development of their own faith traditions. The Center's third component will be a platform for interreligious dialogue between community leaders, clerics, and teachers of the three faiths, who will be invited to study with us and discuss with each other pressing issues of religious interest as guests of the Center. CRIS is a dream that is gradually becoming reality; a dream that makes explicit the critical philosophy I have discussed earlier in the study of religion.

The project sounds terrific. I want now to explore your critical philosophy by asking you about its connection to postmodernism. If I understand it correctly, your project supports the postmodernist posture, especially the so-called linguistic turn. Can you explain how you understand the relationship between postmodernism and Judaism?

That's a large question, but I'll try. I do subscribe to a position rightly categorized as postmodern, but perhaps not in the usual sense of the term. A good point of departure is the frame-dependency in-principle of any criticism, and hence of all rational, reflective action, of which I spoke briefly above. Rational acting and thinking necessarily requires normative commitment. To judge things to be thus and so necessarily requires a vocabulary of appropriate concepts. To judge things to be in any way sufficiently wanting or lacking to require intervention necessarily requires firm commitment to an appropriate system of norms or standards of propriety.

Kant understood this perfectly well, but being the modernist he was—committed to the basic correctness of Newtonian physics as well as to the universality and absoluteness of the moral law—he deemed all experiential knowledge to be governed by a system of timeless and fixed synthetic a priori truths hard-wired into the human psyche, and morality to be grounded universally and absolutely in the very idea of autonomous self-rule. (Only in the third Critique, when dealing with aesthetic judgment, does he seriously consider normative diversity.)

For thinkers like myself, the overthrow of Newtonian physics and the acceptance of ethical diversity have rendered the absolute universalism of Kantian truth and morality by and large obsolete, but not the deeply Kantian idea of the framework dependency of all judgment. Taken together, the neo-Kantianism to which I subscribe is firmly committed to the idea that all human action and thinking is profoundly and necessarily framework-dependent, as well as to that of the all-pervading diversity in principle of normative frameworks governing all realms of human normativity—including ethics, politics, religion, and science.

But now comes the tricky part. The idea of normative diversity and the possibility of framework transitions by no means undermines the obligation we feel toward the norms to which we are at any time committed. In fact, the things for which we are most willing to sacrifice our lives are precisely the things to which our commitment is most blatantly contingent: dedication to our loved ones, faithfulness to our country, our religious obligations, and so forth. Although we are well aware that had the circumstances of our birth and rearing been different, those very commitments would have been radically different, such awareness does not in the least weaken the seriousness of our obligations!

It never occurred to me to associate the postmodernist posture with willingness to die for anything.

Well, as I said, I'm not defending postmodernism as it is usually perceived—namely as a rather silly form of relativism for which the very existence of normative diversity is taken as proof that any judgment is as good as any other. The problem with such an argument is that at the same time that it supposedly acknowledges the framework dependency of judgment, it denies the framework dependency of the very judgment in pronounces! There is no normative framework from within which one can coherently claim that all judgments are equally good (and hence equally worthless). If all judgment is in principle framework-dependent, that type of relativism is profoundly incoherent—not for lack of objective certainty, but for lack of normative basis.

I'm a foundationalist of sorts in insisting that without the foundation of firm normative commitment, we'd have nothing to judge *by* and would, therefore, we be incapable of rationally acting or thinking. But I am a non-foundationalist in insisting that that in order to be considered properly rational, we must somehow be able to exert a critical grip on the very normative foundations of our reasoning, which, as I argued earlier, can be achieved by exposure to trusted and potentially ambivalating normative criticism from without. However, there is no assurance that the normative adjustments we undertake as a result will be in the right direction—in any objective sense of "right." In *The Anxiety of Influence*, Harold Bloom describes the "strong poet" as motivated by a desire to be original and driven by the nightmare of discovering that he's a mere replica. What motivates the critical philosophy to which I subscribe (which I firmly believe I share with Talmudic Judaism's major voice) is a different kind of anxiety: the anxiety, not of influence, but of fixation; a fixation nourished by the comfort zone of agreement and self-assurance; fixation bred by talking to oneself and limiting serious dialogue to one's same-thinking colleagues and cohort. In which case, even when critical, one will be reinforcing one's framework, rather than questioning it. Can we question it directly? No, we can't. But we can sometimes destabilize it a little by exposing it to people with whom we disagree. And once sufficiently destabilized, we can boast having achieved a critical grip on it to rightfully claim to have endorsed the challenged norm or its replacement rationally, not merely because it is an a priori truth.

You advocate the need to subject oneself to constant self-criticism, but what does that mean in terms of religious practice? For example, when one prays in the synagogue where is there room for ambivalence and ambiguity, for ambivalation and self-criticism? Prayer is a conversation that takes for granted that (a) one is heard, (b) one is loved, and (c) one can love in return, and these assumptions do not leave room for self-criticism and ambivalence.

In very extreme situations, criticism of the framework may result in its total abandonment. A physicist or musician may become so fed up with what they're doing as to deem it a waste of time. And the same goes for people who lose their faith and turn their backs on their religion. The kind of normative criticism that interests me in science is not being critical of, say, biology as a vocation, but working for biology by attempting to work against the scientific norms that are constitutive of the way biology is now pursued. A paradigm shift *in* biology is not a paradigm shift *out of* biology.

And the same goes for God-centered forms of religion. Losing one's faith and walking out of the mosque or synagogue is not a framework transition in the sense in which I'm interested. What I'm talking about are framework transitions and norm realignments as a result of criticism from within. Now, the understanding of prayer is not at all as monolithic as you describe it, certainly not in the Talmudic literature.

I was referring to the practice of prayer which requires a certain emotional commitment.

Yes, although, there are some who view prayer as a formal duty devoid of emotional commitment. But even at the level of emotional commitment in prayer, we have seen that Talmudic Judaism presents two very different options or voices: one submissive, one confrontational. The very different religious norms of disposition they describe open up ample space for self-criticism. The plurality of rabbinic positions on prayer is strikingly evident in Lamentations Rabbah, the early Amoraic midrash on the Book of Lamentations. The midrash presents a wonderful mixture of at least three different voices: one voice reads Lamentations as a breast-beating acknowledgment of our responsibility for the destruction and lament of our sinfulness; another, as an angry and pained protest against God's disproportionally heavy and cruel hand; and yet another reads the book as God's own lament of the fate of His people. Once you realize that these voices premise very different theologies, you can appreciate the self-critical, deliberative, and potentially ambivalating nature of Talmudic Judaism, even with regard to standing before God. The difference between the submissive disposition in prayer entailed by a theology of perfection, as premised by the former voice, and that entailed by a theology of imperfection, as premised by the latter two, is enormous. Needless to say, it is the second, confrontational voice that I find both fascinating and an extremely valuable resource in view of modernity's challenges to religion.

Your support of theology of imperfection supports the posture of perplexity.

I know that my reading of the Talmudic literature will sound to many both tendentious and apologetic. But I honestly believe—though would be hard-pressed to prove—that I am not imposing the modern voices to which I concede on the ancient texts except in bringing to the fore opinions that are found there! What I bring to the table are philosophical sensitivities that lend such voices a visibility largely lost on less informed readers, but I don't think that my readings do violence to the text (although they obviously do violence to most traditional ways of reading them).

Since you are supportive of a postmodernist posture, what do you think has been the impact of postmodernism on Jewish practice or Jewish studies? Has postmodernism impacted Judaism and Jewish studies positively? Should Jews welcome and cherish postmodernism?

I thought we had put the postmodernism question to rest.... As I explained, the postmodernism to which I subscribe is very different from the identification of normative diversity with a noncommittal mish-mash according to which "anything goes," as the term is often understood. Such a position flies in the face of everything we cherish in our humanity: our agency, our ability to self-govern and hold ourselves accountable to the norms and ends to which we are committed, our very ability to distinguish right from wrong and true from false as more than matters of mere whim, custom, or taste. Of everything, in other words, that distinguishes us from other animals, no matter how intelligent.

As smart as my cat is, and as powerful as my computer will ever be, they will never ever torture themselves for giving into an impulse they do not approve. Such feats of normative self-governing are exclusively human, and can only be performed by means of norms and standards to which one is committed—not norms we understand, tentatively accept, or hypothesize, but norms to which we are committed and that therefore belong to the framework constitutive of our personal identity and of how we judge ourselves and others. Human beings are reflective by nature, and are hence by nature normatively committed. It is also an empirical fact that the various value systems to which they are committed diverge, at times decisively. This is a relativism that cannot be coherently denied. Where I part company with the flippant and superficial relativism presented all too often as postmodernism is in insisting that knowing that others are committed differently does not detract in the least from the depth and seriousness of our own commitments. Knowing that other people worship other gods, are patriotic citizens of other countries, subscribe to other ideals, and are devoted to other individuals, communities, and vocations as solemnly as I am to mine, may (and, I believe, should) render us more sensitive and considerate of their right to their way of life, but does not (and I believe, cannot) render us less committed to our own, as what all too often goes as postmodernism would have us think.

You just gave us a wonderful formulation of pluralism. Let's explore the problems of pluralism, both for Jews in the diaspora and even more so for Jews in Israel, and also for Israelis who are not Jewish. Why is diversity within Judaism so important to you? How would you justify pluralism within Judaism to people who do not share your view religiously and politically?

I know this is too broad a question, so for starters, can you say more on the concept of tolerance. Is tolerance a necessary point of departure for any pluralistic position?

Let me make one distinction first: there is a difference between plurality and pluralism. And I want to use pluralism in a slightly more specific sense. Plurality is the phenomenon of diversity, in this case, normative diversity; pluralism is the position that values such diversity, that sees it as a desirable blessing. However, to *allow* for diversity and to *favor* it are two different things. Hence the categorical difference between tolerance and pluralism. Tolerance, at the communal level, is willingness to *put up with* the presence of communities dedicated to *objectionable* ways of life. Not all diversity is objectionable, of course. We are fascinated by, and largely welcome, forms of enriching and nonthreatening cultural diversity, like that of cuisine and art. The rabbis even coined a special *berakha* (benediction) to be recited when faced by the teeming multitude (*ukhlosin*) crowding the streets of Jerusalem, with its radical diversity of vocation, taste, and opinion. Tolerance is required, not with regard to forms of life merely different, even radically different from one's own, but with regard to forms of life one finds offensive, when diversity per se outruns itself, as it were. As noted in the second essay in this collection, Avi Sagi has written insightfully on the difference between tolerance and pluralism.

Toleration is the willingness to endure the objectionable, minimally for the sake of peace; a willingness to live and let live without shooting over the fence, as it were. But there are far more benign forms of toleration—liberalism, for instance. The liberal's willingness to endure the objectionable stems from a solemn acknowledgment of the fundamental right of every person to live the life of their choice (insofar as it does not infringe on anyone else's right to live theirs). But even this benign form of toleration lays no intrinsic value on plurality itself, only on the right of others to live as they see fit. Liberals (qua liberals) are tolerant as a matter of principle, but without being compelled to be pluralists.

The Vatican II reforms are remarkable in this respect. The document entitled *Dignitatis Humanae* firmly and movingly acknowledges the fundamental right of every person to his or her religion to the extent that it instructs Christians to proclaim Christ only in ways considerate and respectful of their interlocutors' religious choices. In terms of religious diversity (the irreligious are not officially granted a place at this table), *Dignitatis Humanae* is as liberal as it gets. However, as I explain in detail in the fifth essay in this volume, in the famous document entitled *Nostra Aetate*, the Catholic Church takes a remarkable religiously motivated step beyond liberalism and endorses a pluralist position toward religious diversity—recognizing not only the right of other religions to their beliefs and practices, but urging Christians to value the other religions sufficiently to engage them in religiously meaningful dialogue.

Still, the fascinating pluralist step taken by *Nostra Aetate* beyond the religious liberalism of *Dignitatis Humanae* falls short of full-fledged pluralism. It certainly deems religious diversity a blessing, but only because it views all religions as sufficiently similar to render them viable partners to religiously meaningful dialogue.

When you say "we" or "us" in this context, who do you mean?

In the context of *Nostra Aetate*, the "we" refers to the community of all God-fearing believers, something akin, to what the thirteenth-century

Menachem ha-Meiri referred to as those "bound in the ways of religion" (*gedurim be-darkhei ha-datot*). But I think Judaism offers us more in this respect. My fifth essay in the present volume is an attempt to fashion a Jewish response to *Nostra Aetate* that, unlike the Roman Catholic version, refrains from premising a fundamental sameness of religious content or disposition in favor of a dialogism akin to that of the Talmud, where otherness is valued for its very otherness; i.e., for the kind of challenge it poses. From such a perspective, thoughtful members of other faith traditions, as well as thoughtful nonbelievers, are deemed valuable partners to dialogue, not because of what we have in common, but, on the contrary, because of what we don't have in common! As I noted earlier, over and above the dignity of difference, it is the religious *value* of difference—nay, its religious necessity!—that Talmudic Judaism teaches us.

How diverse is the state of Israel?

Not enough.

In what way it is not sufficiently diverse?

For one thing Israel is seriously impoverished in terms of its *Jewish* religious diversity. It sounds paradoxical, but due to her Orthodox hegemony, Israel is now the only country in the Western world where Jews do not enjoy full religious freedom! Conversions, marriages, and divorces performed by Conservative and Reform rabbis are not recognized by the state, because they are not recognized by the Orthodox rabbinate. (Imagine the British authorities refusing to acknowledge Christian conversions, marriages, and divorces performed by other than Anglican priests!) Unfortunately, the Reform and Conservative movements chose not to partake in the Zionist revolution from the beginning, and, therefore, until very recently, never became an integral part of Israeli society, so their marginality can be understood, though not justified. Had they been partners to the making of the state from the very beginning, there's a good chance that modern Israeli Orthodoxy, would have developed very differently.

Do not forget that the Conservative movement in America has been Zionist from its inception.

Yes, they supported Zionism from afar, cheering it on from the sidelines, but did not participate in the dreaming, planning, and building of Israeli society and state. Conservative and Reform Judaism were not part of Israel's nation-building project, and made their presence known here only relatively recently, far too late to have had the pluralizing effect they should have had on Israeli religiosity. Blinkered by Orthodox-dominated shortsightedness, Conservative and Reform Judaism are not perceived here at all as original and vibrant forms of Judaism, but as inappropriate, watered-down versions of the Orthodoxy they actually know to be true!

In other words, they are viewed as diluted Judaism.

This is how the Orthodoxy in Israel views them—a perception largely accepted by most Israelis. The seriousness and depth of Conservative and Reform Judaism's critique of and alternatives to Orthodoxy, which are well understood and appreciated outside of Israel, especially in the U.S., are little known in Israel; and the seriousness of their rabbinical training, their schools, their services, and communal activity, wholly unappreciated. Conservative and Reform Judaism are, in essence, not only bona fide forms of Jewish religiosity by right, but, in my opinion, are to be valued as the much-needed partners in dialogue to an Israeli Orthodoxy that doesn't even realize they exist.

The problem in Israel is even more complicated because the majority of secular Israelis also don't resonate with the religiosity of American Judaism. They too think that Orthodoxy is the only way of being "truly Jewish." So how do you explain the total failure of liberal forms of Judaism in Israel? The failure to appeal to secular Israelis as well as to be taken seriously by Orthodoxy?

As I said before, the two movements chose not to be here at the time Israel was hammering out her most basic social and cultural arrangements and self-understandings, and they are still not properly here! Israel took form with a deep sense of her Jewish identity and purpose wholly devoid of Conservative and Reform input. Insofar as religion enters her self-identity, and it does so on many levels, it is to Jewish Orthodoxy that she looks exclusively. She boasts a type of profoundly Jewish secular atheism that hardly exists elsewhere, whose Jewish identity is almost wholly exhausted by the civic dimensions of Israeliness. American Jews can hardly resonate with secular Israelis, and conversely, secular Israelis deem American non-Orthodox forms of Judaism an essentially diaspora phenomenon irrelevant to their Israeli existence. This situation, as I said, though understandable, is highly lamentable—especially for Israeli Orthodoxy that has evolved alarmingly unchecked!

Well, Conservative and Reform Judaism are here now. So, how should liberal and progressive Judaism impact Israeli culture?

First and foremost by claiming their full right to religious liberty, and to do so with the firm backing of the two movements worldwide, voiced loud and clear—especially loud! Official Israel can ignore the tiny pockets of Israeli Conservative and Reform Judaism, but she cannot afford to ignore world Jewry. Orthodoxy forms but seven to eight percent of Jews worldwide. The vast majority of Jewish support for Israel—political, moral, and financial—comes from Jews who subscribe to a Judaism that is blatantly discriminated against by the state they support. We're beginning to do a little better now in terms of interreligious conversation, but we still have a very long way to go.

The status of women in Judaism is more complicated than the status of non-Orthodox or nonobservant forms of Judaism. What has been the impact of feminism on Judaism? How do you see Jewish feminism? How important or unimportant is it for you?

Feminism is the great social revolution of the twentieth century. Jewish feminists were right to protest the total absence of women from the entire library of traditional Judaism. There are virtually no Jewish texts written by women until 1900, and very few thereafter. It is not that texts written by women were barred from publication. Texts were not written by women in rabbinic circles, because women were kept by and large illiterate! Although Jewish male society was extraordinarily literate, women members were granted virtually no education to speak of, and therefore, until the emancipation, they contributed nothing to the Jewish bookshelf and nothing to that of rabbinic Judaism, until very, very recently. Female ignorance was meticulously preserved!

But we are now in the midst of a fascinating feminist revolution. The interesting developments in this regard take place today not in liberal and progressive forms of Judaism, where gender equality has long been taken for granted, but within Israel's Orthodox communities. It is heart-warming to witness how women are gaining voice, first of all, in Torah study. Although there is no formal ordination for women in Orthodox circles, which I'm sure will be the next step, there is already a small cadre of highly learned women with real halakhic authority in their communities. More common are women who act as official "pleaders" (barristers) in the rabbinic courts, and it is truly heartwarming to see them arguing thorny halakhic issues on behalf of women before such tribunals. The levels of rabbinic learning available to Orthodox women today is light years away from what it used to be. In most of these frameworks, women still do not study Talmud as do the men, but in several institutions even that curricular gap is gradually being closed.

There have also been remarkable developments with regard to synagogue ritual. There are now Orthodox synagogues that are exceedingly more egalitarian than we could have imagined a decade ago. *Shirah Chadasha* in Jerusalem was the pioneering force in that respect.

But this egalitarian form of Orthodoxy is still small in numbers and that approach is not shared by most Israelis, right?

Indeed, it not shared by the vast majority of Orthodox Israelis and has come under attack from rabbinic quarters. But more and more *Shirah Chadasha* type minyanim are being established in which egalitarianism has become the *religious* norm. There is definitely a movement within traditional society and it is growing very fast relative to how traditional societies are usually apt to move. Something deep is happening in Jewish learning, in Jewish leadership, and even in the traditional Orthodox formations of prayer and ritual.

So, we can be optimistic in that regard?

Yes, but we must not be impatient. Such things take time. Working against deeply ingrained traditional ways of thinking and doing is to work against commitments that are constitutive of one's community's very way of life. And to change a way of life from within, as we have seen, involves a complicated process of exposure to thoughtful and potentially ambivalating criticism, which can prove very tricky indeed. Nonetheless, for a die-hard male Hillelite like myself, the constant challenge of serious committed Jewish voices decidedly different from my own, leveled from across denominational, interreligious, irreligious, or gender divides, is the very lifeline and source of religious vitality.

And how does all that apply to attitudes toward the non-Jews in Israel, be they Arab or Palestinian? Do you see a shift there as well?

At the level of tolerance toward Israel's other religions (Israel I believe, boasts eleven officially recognized religions), I think we're better off than we might seem to be. Ron Kronish's recent *Coexistence & Reconciliation in Israel: Voices for Interreligious Dialogue* paints an impressive picture of the broad range of serious interreligious engagement in Israel today. But in this holy of lands, religion and politics are hopelessly intertwined, and the political problems run deep. I can only speak for the Israeli/Jewish side, of course. For a worryingly growing majority of Israeli Jews, official Israel's exilic siege mentality has all but replaced the values of political Zionism set forth in the Declaration of Independence as the linchpin of Zionist consciousness. The military threats to Israel remain real, though exceedingly less acute than those she faced when the Declaration of Independence was drawn up and signed. Today, however, all opposition to Israeli policy, by friends and foes alike, is conceived almost exclusively in ethno-religious rather than political terms—as forms of perennial anti-Semitism, rather than as solvable issues of state policy. Allowing the battle lines to run between anti-Semite and Jew gives rise to a ghetto-cum-fortress mentality that dramatically shifts attention away from the crucial and delicate tasks of nation-building and statecraft toward an exclusive focus on military force and state economy (as opposed to social welfare). Indeed, if what we are facing is anti-Semitism, pure and simple nation-building and statecraft are rendered superfluous! Such an environment (that as I argued above, feeds at a deeper level on the paradigm of *Shivat Zion*) is detrimental to bond-creating, peace-seeking dialogue across interreligious divides both within and without. Decoupling religion from politics would allow demoting the Israeli-Palestinian conflict from an insurmountable "clash of civilizations" to a heated border dispute. But with the rabbis and mullahs on both sides almost without exception fanning the flames of religious war, normalizing the conflict with a return to political Zionism requires the kind of charisma, vision, and statesmanship that seems at present to be nowhere in sight.

The new book by Amos Oz and his daughter Professor Fania Oz-Salzberger is a very intellectual, text-oriented formulation of secular Zionism that is also very respectful and informed of the Jewish religious tradition.

Yes, the book shows respect to certain voices within the tradition, but the Jewish tradition is more complex than the book conveys. Picking and choosing the voices we like is relatively easy. Changing people's religious commitments grounded in voices we don't like is an entirely different matter—and especially at crucial, revolutionary moments like the Zionist revolution, which require radically new readings of old texts. For political Zionism to be considered a Jewish revolutionary development, rather than a revolution *against* the Jewish tradition, it has to be able minimally to speak Jewish, and the very notions of citizenship, statehood, government, and human rights and liberties deemed halakhically viable. This is far from easy. As I argue in the second of my essays included in this volume, while halakha is reasonably well equipped for tolerance—i.e., for passively putting up with the objectionable for the sake of peace, for instance—it is wholly unequipped for actively assisting sinful behavior. However, that is precisely what the assumption of sovereignty requires. Thus, to take the example I analyze in that essay, former Chief Rabbi of Israel, Bakshi Doron, invoking Maimonides, ruled that since Christianity is a form of idolatry, it is forbidden for Jewish travel agents to cater for Christian pilgrims to the Holy Land. But what would have happened had Israel's Minister of Religious affairs had asked the rabbi the same question? He does not have the privilege of choosing his customers. It is his ministerial responsibility to fund and actively maintain the holy places of all religions, including Christianity, as is that of the Ministry of the Interior to cater for the gay community, etc. Halakha has no answer for questions like these, because it was never seriously thought through in the context of modern statehood and government.

This is a crucial point, because you are saying that in the State of Israel, the state is responsible for actively enabling all forms of expression, both religious and nonreligious.

Of course, as in any modern state. Indeed, the secular educational system is flourishing, as are the churches and mosques of Israel, which are all funded by the government. And yet halakha has yet to acknowledge, let alone bend its mind around, such halakhic anomalies, despite them being the very ABCs of modern sovereignty. In this respect the fundamental rendering of political Zionism in bona fide Jewish terms has not even begun!

Does the Ministry of Education fund activities which are decidedly atheistic?

They do, although I'd use secular, or irreligious, rather than atheistic. The secular youth movements, for instance, will take advantage of the long weekend vacations for their field and camping trips when a major festival falls close to Shabbat. It is the authorities' responsibility not only to fund such blatant halakhic digressions, but make sure that they are performed most effectively! The problem hasn't even registered in halakhic discourse. I know of no serious rabbinic discussion of such issues. There was a flurry of relevant dispute very early in the game, when the then-young Yeshayahu Leibowitz raised several questions related to the Shabbat observance of such state activities that though necessary, could not be halakhically exempted on the basis of *pikuach nefesh* (the principle that Shabbat observance be suspended in life-threatening situations). His examples were the merchant fleet, the banking system, and the diplomatic corps, which couldn't afford the shut down for Shabbat and festivals, but unlike the army and the medical system, could not be halakhically exempted. Despite the intriguing solution he proposed at the time, he got nowhere. The Orthodox Israeli rabbinate proved incapable of even recognizing the need posed by the renewal of Jewish statehood for rethinking Shabbat observance in other than personal and communal terms. The crucially important work of rethinking and adjusting, expanding and enriching Judaism's basic vocabulary—theologically, halakhically, and liturgically—to accommodate the challenges of modern Jewish sovereignty and statehood, has hardly begun, and Israeli Orthodoxy shows no sign of doing so. The Haredi, ultra-Orthodox community remains largely indifferent to the problem, while the Religious Zionist rabbis are zealously busy conceptualizing Israeli statehood in worryingly premodern terms. Because the halakhic issues of sovereignty are left unattended, and the deep ethos of modern democracy is deemed seemingly untranslatable into Jewish terms, the rift between religious and political Zionism is swiftly widening. What is at stake is not the relationship between state and religion, but the very possibility of squaring religious faithfulness with modern statehood.

But if you frame the issue in that way, aren't you saying that we have to think about the role of the non-normative person in the context of a normative system that does not validate that person?

I certainly am. This is precisely why I find Talmudic pluralism such an important latter-day resource for facing up to these grave challenges. If your ideal is diversity rather than sameness—real normative diversity, that is—then as long as a form of life is thoughtful, it is of value for the challenge it poses, as well as for what it can teach us, regardless of its non-normativity from our own point of view.

Obviously, we agree that education is very important to the developing of the society that will be at least in potential pluralistic.

Obviously.

What's the role of Jewish studies in the educational mission that you see for Jews, in Israel and in the diaspora?

I can think of three major educational roles for Jewish studies pertaining to our three main concentric circles of engagement. The first steers close to what Fania and Amos Oz's book exemplifies: namely, the deep sense of belonging we experience when we identify voices close to our own within Judaism's textual history. The realization through learning that we don't have to speak for the tradition, because the tradition, as it were, speaks for us, lends historical and cultural depth to our sense of who we are. For it allows us to become an integral part of the whole without having to sacrifice anything of our self identity. The more one studies Judaism the greater the sense one has of it being a kaleidoscopically diverse whole in which one's unique voice can be at home; more like a shared language—in which almost anything can be said and maintained—than a shared system of thought.

The second educational role I see for Jewish studies builds on the first. And here the Talmudic period is central. What a Talmudic perspective adds to realizing the kaleidoscopic diversity of Jewish thought and practice and one's place within it, is its enriching value. It enables us not only to feel welcome as participants in the tradition's great spectrum of voices, but to value and view as a blessing—as opposed to merely tolerate—the trying presence of voices not merely different, but actually opposed to our own. Viewed through a Talmudic lens, the spectrum of contrasting Jewish voices is transformed from a grid within which to position oneself, to an all-enriching conversation in which one is invited to partake.

However, Jewish studies has more to offer than a haven and study hall for its Jewish participants. The Talmudic literature paints a largely fantasized picture of keen Roman interest in the rabbinic world and of a reciprocal rabbinic willingness to engage in such dialogue. Today this is no longer a fantasy. At least since the late eighteenth century, Jewish philosophy, religious and secular, has developed in keen, and explicit dialogue, with a broad spectrum of especially modern European religious and secular philosophy. And since the more recent establishment of religious studies as a free-standing academic discipline, and more so with the impressive proliferation of interfaith dialogue, especially in the wake of the Vatican II, there has been a surge of interest in Judaism worldwide. All aspects of the Jewish tradition, biblical and rabbinic, philosophical and mystical, theological and practical, liturgical and legal, are now studied in earnest by non-Jews in both academic and nonacademic contexts.

What the Talmudic literature could only imagine has now become a reality that far transcends the Talmud's wildest dreams. But dreams aside, what Talmudic Judaism still has to offer us is a vivid picture of the religious meaningfulness of such engagement. Paradoxically, the very literature that signaled Judaism's retreat from the world of state politics, more than any other, can be seen, when studied properly, to pave the way to reentering the world by engaging it philosophically, culturally, and interreligiously.

That sounds to me a very optimistic approach to the future for Jewish studies. Would you say we are today where we were in the 1970s and 1980s in terms of the status of Jewish studies? I don't think so, because I see a decline in the status of Jewish studies.

My impression is different, but it is only an impression. But what you asked me about was the educational role I saw for Jewish studies. I can't speak of what the future bears, only of the potential promise of the present and the recent, and the not so recent past. Political Zionism's revolutionary dream of benign, modern Jewish statehood is worryingly threatened by a corroding combination of messianic and exilic Israeli mentalities that cast a dark shadow over Israel's very ability to realize her initial dream. Here too, as with Jewish studies, I cannot say what the future holds, only that in both cases, it is largely up to us.

What do you see as the challenges to Jewish existence today?

In my opinion the challenges to the very existence of Jewishness are not external, but internal. We are no longer in real danger of physical annihilation, despite the rise of anti-Semitism proper (as opposed to, and all too often confused with opposition to Israeli policy) and the military threats faced by Israel. But we are in real danger of Judaism losing its relevance and vitality for Jews. The greatest challenges we face in this respect are the indifference of many Jews to their Jewishness, the temptations of Haredi withdrawal and stagnation, and the lure of delusionary messianic denial of hardcore religious Zionism. But we're living in a dramatically changing world. Until very recently, a mixed marriage was synonymous with marrying out—i.e., a decision on behalf of the Jewish party to turn his or her back on their Jewishness and walk away. But we now face a new reality, insofar as very many mixed couples in Reform communities remain proud bona fide members of the community as a matter of choice.

Do you refer to spouses who function as members of the community but without conversion?

Yes, I do. The notion of belonging to the Jewish people without necessarily becoming Jewish, brings me back to the Book of Ruth, of course. Ruth joins the Jewish people without undergoing conversion (a concept foreign to biblical Israel). She becomes part and parcel of the Jewish community of Bethlehem-Judah while referred to throughout as "Ruth the Moabite" by the biblical text. Just as in Ruth, one can be Jewish today without becoming Jewish, as it were—a reality that many Jews still consider wholly oxymoronic.

You suggest, then, that Judaism is thriving as a religious practice that appeals to non-Jews in America.

I am, but Judaism cannot afford to stagnate. The key to Jewish flourishing, teaches the Talmud, is exposure and attention to voices external to the community's in-house deliberations. The presence of non-Jewish members in Jewish communities can be an important asset in that respect, but only against the backdrop of serious Jewish learning and deliberation. Hence the important threefold role of Jewish studies, of which we spoke before. Jewish practice may be a necessary condition for the future of Judaism, but it is not a sufficient one. Practice delineates communal boundaries but cannot ensure their vitality and flourishing. Here Jewish *learning* is absolutely crucial, in all its three dimensions. For it is by serious study of Judaism's textual history motivated by the urgency of the problems we face, far more than by participating in Jewish rituals, that we can hope to attain sufficient levels of confidence, knowledge, and pride in our tradition, to dialogically engage ourselves and the world in the spirit of our Talmudic heritage. This is my great dream.

This is an important message to share with Jews and non-Jews. Thank you very much for taking the time to interview and to participate in the Library of Contemporary Jewish Philosophers.

Thank you for including me.

SELECT BIBLIOGRAPHY

Books and Monographs

1. (Edited with S. J. Schaffer) *William Whewell: A Composite Portrait.* Oxford: Clarendon Press, 1991.
2. *William Whewell: Philosopher of Science.* Oxford: Clarendon Press, 1991.
3. *To Know Wisdom: Science, Rationality, and Torah-Study.* Tel Aviv: Hakibbutz Hameuchad, 1994 (Hebrew).
4. *Rational Rabbis: Science and Talmudic Culture.* Bloomington: Indiana University Press, 1997.
5. (With Y. Benbaji) *The View from Within: Normativity and the Limits of Self-Criticism.* South Bend, IN: University of Notre Dame Press, 2011.
6. *Creatively Undecided: Toward a History and Philosophy of Scientific Agency.* Chicago and London: The University of Chicago Press, forthcoming.
7. *The Rabbis' Dispute of Religiosity: A Study of Talmudic Covenantal Theology* (in preparation) (Hebrew).

Journal Articles

8. "Hempel's Ravens, The Natural Classification of Hypotheses and the Growth of Knowledge." *Erkenntnis* 21 (1984): 45–62.
9. "Necessary and Contingent Truth in William Whewell's Antithetical Theory of Knowledge." *Studies in History and Philosophy of Science* 16 (1985): 275–314.
10. "Whewell's Consilience of Inductions—An Evaluation." *Philosophy of Science* 52 (1985): 239–55.
11. "A Physicists' Philosopher: James Clerk Maxwell on Mathematical Physics." *Journal of Statistical Physics* 51 (1988): 309–19.
12. "The Problematic History of Explaining Science." *Alpayim* 1 (1989): 194–208 (Hebrew).
13. "Being, Becoming, Creation and the Crisis of Baconianism in the Nineteenth Century." *Alpayim* 4 (1991): 224–32 (Hebrew).
14. (With N. J. Efron) "Science Naturalized, Science Denatured: Reflections on Ronald Giere's *Explaining Science.*" *History and Philosophy of the Life Sciences* 13 (1991): 187–221.

15. “Ein Blick vorwärts in die Vergangenheit: Ein Fall für den historischen Nominalismus” (Looking Forward to the Past: A Case for Historical Nominalism). *Deutsche Zeitschrift für Philosophie* 40 (1992): 1279–94.
16. “The Emergency Which Has Arrived: The Problematic History of 19th Century British Algebra—A Programmatic Outline.” *The British Journal for the History of Science* 27 (1994): 247–76.
17. “Towards a Rational Theory of Progress.” *Synthese* 99 (1994): 277–304.
18. “Trouble-Shooting Creativity.” *History and Philosophy of the Life Sciences* 16 (1994): 141–53.
19. (With H. Shapira) “The Debates between the Houses of Shammai and Hillel: The Meta-Halakhic Issue.” *Iyyunei Mishpat* (*The Tel Aviv University Law Review*) 22, no. 3 (1998): 461–97 (Hebrew).
20. “The Making of Peacock’s *Treatise on Algebra*: A Case of Creative Indecision.” *Archive for History of Exact Science* 54 (1999): 137–79.
21. “Scientific Dilemmas: Splits and Doubles in Early Victorian Science.” *Zemanim* 66 (1999): 77–90 (Hebrew).
22. “When Philosophers Encounter Texts.” *Theory and Criticism* 15 (1999): 19–24 (Hebrew).
23. (With N. J. Efron) “Astronomic Exegesis: Interpretation of the Heavens by Early Modern Jews.” *Osiris* 16 (2001): 72–87.
24. “The Covenant of Confrontation.” *Alpayim* 22 (2001): 9–52 (Hebrew).
25. “A Modest Proposal: Toward a Religious Politics of Epistemic Humility.” *Journal of Human Rights* 2, no. 1 (2003): 49–64.
26. (With Y. Benbaji) “Through Thick and the Thin: A New Defense of Cultural Relativism.” *Southern Journal of Philosophy* 42, no. 1 (2004): 1–24.
27. “Canon, controverse et réforme; une réflexion sur l’autre voix du judaïsme talmudique” (Canon, Controversy and Reform: A Reflection on the Other Voice of Talmudic Judaism). *Les Cahiers du judaisme* 18 (2005): 61–75.
28. (With Y. Benbaji) “Factuality without Realism: Normativity and the Davidsonian Approach to Meaning.” *Southern Journal of Philosophy* 43, no. 4 (2005): 505–30.
29. “Taking the Bavli Seriously.” *Diné Yisrael: Studies in Halacha and Jewish Law* 23 (2005): 75–91.
30. “*Brakhot* 19b: The Bavli’s Paradigm of Confrontational Discourse.” *Journal of Textual Reasoning* 4, no. 2 (March 2006). http://etext.virginia.edu/journals/tr/volume4/number2/TR04_02_a01.html.
31. “Criticism, Interpretation and Canon: A Reply of Sorts.” *Journal of Textual Reasoning* 4, no. 2 (March 2006). http://etext.virginia.edu/journals/tr/volume4/number2/TR04_02e05.html.

32. "*Rational Rabbis*: Its Project and Argument." *Journal of Textual Reasoning* 4, no. 2 (March 2006). http://etext.virginia.edu/journals/tr/volume4/number2/TR04_02_e01.html.
33. "Legislation and Legal Norm Development in Jewish Law." *Periodica de re Canonica* 96 (2007): 401–17.
34. "Taking the Linguistic Turn Seriously." *The European Legacy* 13, no. 5 (2008): 605–22.
35. "The Ethics of Memory: Religious Amnesia and Social Justice." *Havruta* (2010): 42–49.
36. "Toward a History and Philosophy of Scientific Agency." *The Monist* 93, no. 4 (October 2010): 518–44.
37. "How and Why I Write History of Science." *Science in Context* 26, no. 4 (2013): 573–85.
38. "Science, Religion and Rationality: A Neo-Hegelian Approach." *Toronto Journal of Theology* 29, no. 2 (2013): 319–36.
39. "Babbage's Two Lives." *British Journal for the History of Science* 47, no. 1 (2014): 95–118.
40. "Criticism from Within *versus* Criticism from Without—My Debt to Yemima." *Iyyun—The Jerusalem Philosophical Quarterly* 64 (April 2015): 160–71 (Hebrew).
41. "Deciding by Argument *versus* Proving by Miracle: The Myth-History of Talmudic Judaism's Coming of Age." *Toronto Journal of Theology* (forthcoming).

Book Chapters

42. Introductory essay to the Hebrew edition of Francis Bacon's *New Atlantis*, 5–20. Tel Aviv: Hakibbutz Hameuchad, 1986.
43. "Il collegamento dei fatti: la nozione non necessaria di verità empirica di William Whewell" (The Colligation of Facts: Whewell's Unnecessary Notion of Empirical Truth). In *L'epistemologia di Cambridge (1850–1950)*, edited by R. Simili, 39–63. Bologna: Societa Editrice il Mulino, 1987.
44. "Antithetical Knowledge." In *William Whewell: A Composite Portrait*, edited by M. Fisch and S. J. Schaffer, 289–309. Oxford: Clarendon Press, 1991.
45. "Learning from Experience." In *Probability and Rationality: Studies on L. Jonathan Cohen's Philosophy of Science*, edited by E. Eells and T. Maruszewski, 65–79. Poznan Studies in the Philosophy of the Sciences and the Humanities 21. Amsterdam and Atlanta: Rodopi, 1991.

46. "A Philosopher's Coming of Age—A Study in Erotetic Intellectual History." In *William Whewell: A Composite Portrait*, edited by M. Fisch and S. J. Schaffer, 31–66. Oxford: Clarendon Press, 1991.
47. "The Perpetual Covenant of Jewish Learning." In *Summoning: Ideas of the Covenant and Interpretive Theory*, edited by E. Spolsky, 91–114. Albany: SUNY Press, 1993.
48. " 'And they shall keep the way of the Lord, to do justice and judgment'—Faith, Knowledge and the Obligatory Nature of the Jewish Religion." In *Yeshayahu Leibowitz—His World and Philosophy*, edited by A. Sagi, 109–20. Jerusalem: Keter, 1995 (Hebrew).
49. "Ecclesiastes (Qohelet) in Context—A Study of Wisdom as Constructive Scepticism." In *Critical Rationalism, The Social Sciences and the Humanities: Essays for Joseph Agassi*, vol. 2, edited by I. C. Jarvie and N. Laor, 167–87. Dordrecht, Holland: D. Reidel, 1995.
50. "Leibowitz, the Shammaites and the Limits of Halkhic Criticism: An Introduction to State Law." In *A Good Eye: Dialogue and Polemic in Jewish Culture*, edited by N. Ilan, 603–27. Tel Aviv: Hakibbutz Hameuchad, 1999 (Hebrew).
51. "Living Dangerously." In *The Jewish Political Tradition*, vol. 1: *Authority*, edited by M. Walzer, M. Lorberbaum, N. Zohar, and Y. Lorberbaum, 328–33. New Haven: Yale University Press, 2000.
52. "From Two to One and Back: The Creation Stories and David Hartman's Covenantal Anthropology." In *Renewing Jewish Commitment: The Work and Thought of David Hartman*, edited by A. Sagi and Z. Zohar, 659–93. Tel Aviv: Hakibbutz Hameuchad, 2001 (Hebrew).
53. "Ruling Others: The Halakhic Challenge of the Renewal of Jewish Sovereignty." In *The Other Within and Without*, edited by M. Ben-Sasson and H. Deutsch, 225–59, 495–502. Jerusalem: Yedioth Aharonot Press, 2001 (Hebrew).
54. "Scientific Values." In *Crossroads: Values and Education in Israeli Society*, edited by J. Cohen, Y. Iram, S. Scolnicov, and E. Schachter, 69–90. Jerusalem: Ministry of Education, 2001 (Hebrew).
55. "Ein bescheidener Vorschlag: Auf dem Weg zu einer religiosen Politik der epistemischen Demut." In *Die religiosen Wurzeln der Toleranz*, edited by Ch. Schwöbel and D. C. v. Tippelskirch, 131–59. Freiburg im Breisgau: Verlag Herder, 2002.

56. "Ethical Diversity, Tolerance, and the Problem of Sovereignty: A Jewish Perspective." In *The Many and the One: Religious and Secular Perspectives on Ethical Pluralism in the Modern World*, edited by T. B. Strong and R. Madsen, 195–218. Princeton, NJ: Princeton University Press, 2003. Republished as "Diversity, Tolerance, Sovereignty." In *Law, Politics, and Morality in Judaism*, edited by M. Walzer, 96–120. Princeton, NJ: Princeton University Press, 2006.
57. "Rationality, Exegesis and Canon." In *To and Fro: Philosophy's Many-Sidedness*, edited by N. Naaman-Zauderer and Y. Senderowicz, 215–44. Tel Aviv: Mifalim Universitaim, 2003 (Hebrew).
58. "Introduction" to H. Fisch, *Who Knows One?—An Essay in Autobiography*, 7–23. Ramat Gan: Bar-Ilan University Press, 2004.
59. "The Crisis of Modern European Science and Religion and the Jewish Debate of Religiosity." In *On Faith: Studies in the Concept of Faith and its History in the Jewish Tradition*, edited by M. Halbertal, D. Kurzweil, and A. Sagi, 512–33, 693–95. Jerusalem: Keter Publishing House, 2005 (Hebrew).
60. "Science, Creativity, and Criticism: Inquiry and Science Education—An Interview with Professor Menachem Fisch." In *Learning by Inquiry: An Ongoing Challenge*, edited by A. Zohar, 85–106. Jerusalem: Magnes, 2006 (Hebrew).
61. "Forced Readings and Binding Texts: The Amoraic *Ukimta* and the Philosophy of Halakha." In *New Studies in the Philosophy of Halakha*, edited by A. Ravitzki and A. Rosenak, 311–44. Jerusalem: Van Leer Institute and Magnes Press, 2008 (Hebrew).
62. "Judaism and the Religious Crisis of Modern Science." In *Nature and Scripture in the Abrahamic Religions: 1700–Present*, vol. 2, edited by J. M. van der Meer and S. Mandelbrote, 525–67. Brill's Series in Church History 37. Leiden: Brill, 2008.
63. (With Peter Pettit) "Religious Self-Critique in the Trusted Presence of Others." In *Voices for Interreligious Reconciliation in Israel*, edited by R. Kronnish, 202–13. New York: Paulist Press, 2015.
64. "Judaism and the Religious Value of Diversity and Dialogue: Drafting a Jewish Response to *Nostra Aetate*." In *Diversität—Differenz—Dialogoizität: Religion in Pluralen Kontexten*, edited by S. Alkier, M. Schneider, and Ch. Wiese. Berlin and Boston: de Gruyter, forthcoming.

Work Published in Proceedings of Scientific Meetings

65. "Humble Reasoners: Science and the Jewish Covenants of Learning." *Metanexus Monthly: The Online Forum on Religion and Science* (August 2002) (Parts 1 and 2).
66. Presentation (pp. 65–70) and extensive dialogue participation in *Modest Claims: Dialogues on Toleration and Tradition*, edited by A. B. Seligman. South Bend, IN: University of Notre Dame Press, 2004.

Items in Encyclopedias and Dictionaries

67. "William Whewell." *Routledge Encyclopedia of Philosophy*, vol. 10, edited by E. Craig, 709–11. London and New York: Routledge, 1998.
68. "George Peacock." *The Dictionary of Nineteenth-Century British Scientists*, vol. 3, edited by B. Lightman, 1555–62. London and New York: Thoemmes Continuum Press, 2004.

Reviews

69. (With N. J. Efron) Review of *Scientific Genius: A Psychology of Science*, by D. K. Simonton. *History and Philosophy of the Life Science* 13 (1991): 313–14.
70. Review of *Defining Science: William Whewell, Natural Knowledge, and Public Debate in Early Victorian England*, by R. Yeo. *Isis* (1994): 706–7.